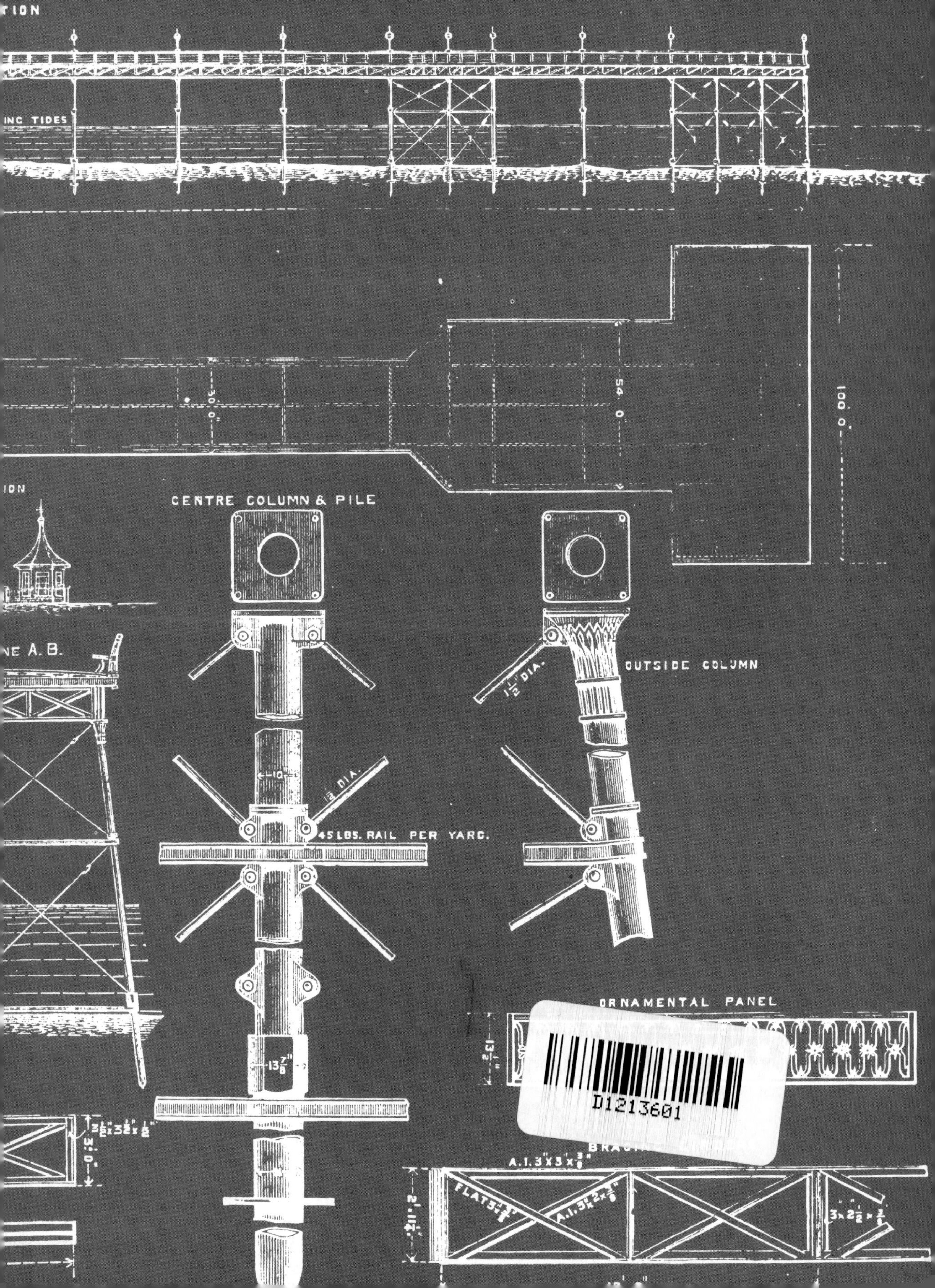

ING TIDES

ION

CENTRE COLUMN & PILE

NE A.B.

OUTSIDE COLUMN

1½ DIA.

410

¾ DIA.

45 LBS. RAIL PER YARD.

13⅞"

13½"

ORNAMENTAL PANEL

3½x3½x½"

A.I. 3"X3"X⅜"

2'.11½"

FLAT 3"x⅜"

A.I. 3½ x 2½ x ⅜"

3 x 2½ x ½"

30.0

54.0

100.0

SEASIDE PIERS

SEASIDE PIERS

Simon H. Adamson

B. T. Batsford Ltd, London
in association with the Victorian Society

'To Margaret'

ACKNOWLEDGEMENT

The Author and Publishers wish to thank the following for permission to reproduce the illustrations appearing in this book: Frith & Co. Ltd. for the frontispiece; Chief Librarian, East Sussex County Libraries for figs 1, 21, 27; J. S. Gray for fig. 2; Herne Bay Records Society for figs 5, 13, 42; *Illustrated London News* for figs 7, 11, 24; West Sussex Library Service for figs 8, 18; Lancashire County Library for figs 9, 19, 28; Head Wrightson & Co. Ltd. for figs 12, 14, 15, 43, 53; Margate Public Library for fig. 23; Valentines of Dundee Ltd. for fig. 33; *Kent Messenger* for fig. 34; K. A. Falconer, CBA Industrial Monuments Survey for figs 35, 36, 51, 63, 65, 67, 68, 73, 75, 76, 78, 80, 81; *The Engineer* for figs 37, 40, 41, 52, 54 and endpaper; *Engineering* for figs 38, 56; Mrs J. Holmes for fig. 39; Sir William Halcrow & Partners and Dover District Council for fig. 44; *South Avon Mercury* for fig. 59; Kent Fire Brigade for fig. 60; Hampshire Fire Brigade for fig. 61; May Gurney & Co. Ltd. for fig. 66. All illustrations not otherwise credited are by the author or from his collection.

First published 1977
Reprinted 1983

© Simon H. Adamson, 1977
© Anthony Dale, 1977 (Chapter 5)

ISBN 0 7134 0242 3
Filmset in 10/13pt 'Monophoto' Century Schoolbook
by Tradespool Ltd, Frome, Somerset
Printed in Great Britain by
The Anchor Press Ltd, Tiptree
for the publishers B. T. Batsford Ltd
4 Fitzhardinge Street, London W1H 0AH

CONTENTS

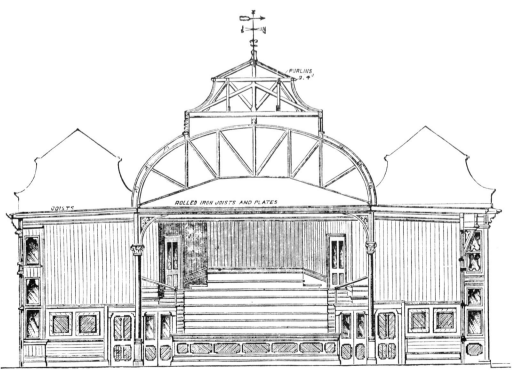

PREFACE

This book could not have been written without the kind and willing assistance of a large number of people. It had its origins in a dissertation on the engineering aspect of piers written for the University of Southampton in 1974, and I am grateful to Professor P. B. Morice, who introduced me to this fascinating subject, and the university staff who assisted in the preparation of the work. Keith Falconer, of the Industrial Monuments Survey, freely gave valuable advice and assistance during the preparation of the later Victorian Society report on piers. Numerous librarians, archivists, local authority engineers, consulting engineers, contractors and pier operators have kindly provided information and photographs, and I am only sorry not all this data could be included. I am very grateful to those organisations who kindly gave permission for illustrations to be reproduced. Such pier researchers and preservationists as Margaret Wilson, Frank Benatt and John Lloyd have all been most helpful, and Graham Pike rendered valuable assistance with the photography. Anthony Dale's informative chapter on pier architecture is based on his experience as historic monuments inspector for the Department of the Environment, and it was ultimately on his recommendations that notable piers have been given 'listed' status. The gazetteer is reproduced from the Victorian Society report, with the Society's permission.

Most of all, however, it was Margaret Nunn who saw all three stages safely into print, and without whom the work could never have come to fruition. Thank you.
Southampton
May 1976

FOREWORD

by Lord Asa Briggs

This study of piers has a double interest. Social historians are now turning increasingly from the world of work to the world of leisure, a sign, perhaps, of changing times; and piers have been favourite places of leisure and pleasure. At the same time, growing public concern for distinctive features in our urban environment has been quickened by the sense that many of these distinctive features are in jeopardy; and Chapter Six of this book exposes the dangers, very real ones, to those piers which have survived. It would be sad if the study of piers became simply a study in non-industrial archaeology. They have too many human possibilities in the future to be left entirely either to archaeologists or to historians.

'Funland' was a late-Victorian conception, and by then piers, like promenades and winter gardens, were signs of seaside progress. Every local history of a seaside resort where a pier or, better still, piers were built includes the date of the first opening as a landmark date in the local chronology. Mr Adamson has assembled a mass of scattered local material, putting it into order and illustrating it, as all good social historians should, with every kind of visual material. But once opened, piers changed. New elements were added and some old features disappeared. There is far more to be written both about the economics of the process and its social driving force. (Compare the building of marinas since the 1950s.) The outline however, is now available: it is useful and illuminating.

The Victorian Society has added greatly to our knowledge while influencing our tastes and values. It is not surprising, therefore, that it is associated with this study. The Society has been interested in Town Halls also—another expression of Victorian culture—and it is in Town Halls or their twentieth-century progeny that the key decisions are being taken, often terribly slowly, about what to conserve from the Victorian past and what to adapt or transform. The Victorians liked pressure politics, and it is through pressure that decisions can be speeded up before they come too late. Local amenity societies can do much but they need national support also. The seaside belongs to us all. We may not all respond to the vision behind one of Mr Adamson's quotations—'it will be necessary to alter the map of England, and represent it as a huge creature of the porcupine type, with gigantic piers instead of quills'—but we do not want to see the last great seaside features of this island disappear. We have ruined huge areas of our coastline first in Britain and then in the Mediterranean during the course of this century, and if we are to rectify the account we must get all the answers right for the next twenty years.

Asa Briggs
Worcester College, Oxford

ONE

INTRODUCTION

Strong winds had been blowing throughout the morning of Saturday, 20 May 1899, along much of the South Coast. In the Sussex seaside resort of Brighton, powerful waves broke on the beach below the promenade, and gusts tugged at the clothing of the few hardy souls who braved the weather to attempt a stroll along the front. They may well have been surprised to see a long procession forming on the promenade. A military band, the Mayor and Mayoress of Brighton, council officials, members of the Town Council, and numerous local notables had assembled on that cheerless day to mark an important event in the development of the town—the opening of the Brighton Marine Palace and Pier.

In truth, they were not so much opening the entire pier as marking a major step in its progress. Most of the pier was finished; the final stage was the erection of the large platform at the seaward end and the pavilion which was to surmount it. The ceremony was primarily designed to mark the start of work on this *pièce de résistance*.

Just after midday the procession started on to the pier. The wind, which made the decorative flags and pennants fly so proudly, also slowed the progress of the party and the ladies experienced not a little trouble with their voluminous dresses. The gentlemen, wearing the top hats so obviously required on such an important occasion, were also hard pressed to keep their dignity. Even the scarlet-clad bandsmen of the Brighton Rifles were unable to walk upright, and had to move with heads bowed into the wind. The Mayoral party moved as quickly as they could to the end of the pier, where the ceremony was to take place. They paused momentarily at the mid-point of the pier to take down a rope as the first token of the opening, and the procession eventually stopped a few yards short of the end of the 1,400ft pier. Further south, the deck ended and there were just a few piles and girders separating the party from the cold grey waters of the English Channel.

A steam crane, its jib towering above the figures on the deck, held up an iron column a few feet above a base plate. The members of the party gathered round this apparatus and Mr John MacMillan, Chairman of the Pier Company, made a short speech, which the audience must have had some difficulty in hearing. The highlight of the ceremony followed. Mr Howard, under whose supervision the pier had been built, handed the Mayoress, Mrs Hawkes, a silver spanner suitably engraved.

THE THREE BRIGHTON
PIERS SPAN THE PIER
BUILDING ERA

1 *above* Chain Pier
(1822–3), c. 1890, of
unusual suspension
design, was destroyed
in a storm in 1896

2 *left* West Pier
(1863–6), c. 1868. one
of the first of the new
wave of pleasure
piers. It is currently
under threat of
demolition.

The column was lowered on to its receiving bolts, a nut was affixed to one, and Mrs Hawkes, wielding her silver spanner, tightened the joint. The assembled party raised three hearty cheers, the band played the opening bars of the National Anthem, the Mayoress was thanked for her presence, and those assembled hurried back to the relative calm of the entrance.

One further ceremony had yet to be performed, however. The Mayor and Mayoress each passed through a turnstile to re-enter the pier, Mr Howard and Mr MacMillan took their tolls, and the pier was thrown open to the public, from 9 a.m. to 9 p.m. at 2d. a head. The invited company, meanwhile, had retired to the Royal Pavilion for luncheon. Speeches were made, toasts were proposed, and the whole occasion was adjudged a great success.

It was fitting that Brighton should have one of the last, and one of the most sophisticated, of seaside piers. The town had been in the van of pier building in the early 1800s, and had epitomised the growth of the seaside resort, the roots of which lay in that eighteenth-century social pursuit, hypochondria.

Early beginnings

Hypochondria was a popular pastime in the upper echelons of British society in 1800, and an industry had grown up to cater for the wealthy sufferers of real or imaginary illnesses. Harrogate, Boston Spa, Tunbridge Wells and, above all, Bath had prospered on the curative powers of their mineral springs. Bath even rivalled London as a social and cultural centre. Libraries, ballrooms, crescents of fine houses and broad avenues were provided for the use and entertainment of visitors to the town. But the spa towns' popularity was soon to be challenged by another prospective panacea for health-seekers—the seaside watering place.

Sir John Floyer had expounded the virtues of sea water in the 1750s. Drinking the stuff, as well as bathing in it, could, it was thought, cure all manner of complaints. When, in addition, a mineral spring was discovered on the beach at Scarborough, the beginnings of the 'seaside watering places' were formed. Previously, coastal communities had grown up as fishing villages, ports, or naval bases. A large proportion of the population seldom, if ever, saw the sea. Those who knew it tended to regard it as unfriendly, dangerous and responsible for a rather

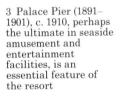

3 Palace Pier (1891–1901), c. 1910, perhaps the ultimate in seaside amusement and entertainment facilities, is an essential feature of the resort

coarse breed of seaman which made its living from it. Certainly the sea was a useful provider of food, it was responsible for much of the country's trade, and it constituted a useful means of defence, but the idea of anyone going to the coast for pleasure or amusement purposes was at one time laughable. Slowly, however, visitors began to forsake the comfortable well-established spas to sample the delights of the seaside. By the early years of the nineteenth century, such places as Scarborough, Weymouth and Margate were accepted as watering places.

Getting to these places, however, presented a problem. For some time, the choice of transport lay between sailing passenger vessels and stage coaches. The former were slow and notoriously unreliable (the London–Margate trip could take anything from nine hours to three days depending on the weather), although they had the advantage of being relatively cheap. Coach travel was faster and more reliable, but more expensive and often highly uncomfortable. These factors deterred many would-be visitors, but the introduction of steam passenger vessels around 1815 swiftly changed the situation. Reliable, comfortable and fast, these vessels had the effect of dramatically increasing the number of travellers to the 'coastal resorts' (as they became known). Often however, the real adventure started when the boats arrived at their destinations, since passengers experienced difficulty in getting ashore. Sometimes, as at Brighton, passengers were taken ashore in small boats, and their luggage was ferried to the beach on rafts. In the case of travellers from France, horses and vehicles often had to be landed, and even a slight swell could be hazardous. The beach was often littered with goods, indicating an accident or carelessness at some stage during the unloading process.

Visitors to the Isle of Wight were in a similar predicament. At Ryde, the principal entry point for the island, travellers faced an uncomfortable ride to the beach

EARLY LANDING
PIERS

4 *left* Ryde Pier,
c. 1840, built 1813–14
to facilitate passenger
landings for the Isle
of Wight. It exists
today in a greatly
modified form

5 *above* Herne Bay
Pier (1831–2) in 1841.
This ambitious pier
was erected to cater
for sea-borne visitors,
but was replaced by a
smaller, pleasure, pier
in the 1870s

on the back of a porter, where at low tide there was a half mile walk to the town over the wet sand. The town fathers of Ryde, as of many other such towns, saw these arrangements as a hindrance to the development of the town. Clearly, they could not expect the number of wealthy visitors to increase (and it was, after all, only the wealthy who had the time and money to indulge in such travel) if these primitive facilities continued. Something had to be done to allow travellers to disembark from vessels in safety and comfortably, at most, if not all, states of the tide. A landing pier was accordingly planned.

Ryde's pier was built in 1813–14 and extended 1,250ft into the sea. It was an immediate success and subsequently more than doubled in length as traffic increased. Ryde's example was quickly followed elsewhere. At Brighton, the famous Chain Pier was erected in 1822–3 to act as a landing stage there. Herne Bay followed Walton-on-the-Naze and Southend in providing landing piers in the late 1820s and early 1830s.

At Margate, increasing steamboat traffic highlighted the shortcomings of the existing harbour facilities, which were inaccessible at low tide. The number of passengers using the port had increased from some 17,000 in 1812–13 to almost 44,000 eight years later. The Margate Pier and Harbour Company therefore erected the so-called Jarvis Landing Place in 1824. This was an 1,100ft wooden jetty which was available only at low water and half tides; at high water it was completely submerged. The structure rose slightly towards the seaward end, and it was not uncommon for unfortunate strollers to be cut off by the rising tide as they enjoyed the sea breezes out in the harbour. The result was a lucrative business for the local seamen who carried the marooned visitors ashore, and amusement for the spectators safe on dry land.

By the 1840s there were a number of well-established resorts, located throughout England and Wales. For example, Margate was joined by such places as Herne Bay, Brighton, Worthing and Bognor in serving the needs of the metropolis and nearby areas. On the Yorkshire coast, Scarborough and Bridlington catered for the wealthy mill-owners and woollen merchants of the county. Several had prospered mainly because of royal patronage—Weymouth was a case in point. It should not be imagined that every little fishing village had, in the course of some fifty years, grown into a prosperous resort with fine buildings and wealthy visitors; but a number had, and some of these featured piers.

These early structures were primarily functional; they admirably did the job for which they were provided. It soon became apparent, however, that many people were visiting these structures for purposes unconnected with travel or trade.

In the 1840s and 1850s the emphasis on the piers' functions was changing. In the early days, pier promoters had sold their companies' shares on the basis of the return expected from tolls charged on passengers and goods landed. Soon the word 'promenade' appeared in the prospectuses, and the possible profit to be made from casual strollers was used as an additional selling point to potential shareholders. Precisely why piers had this peculiar fascination for strollers is difficult to define. Possibly Brettell, in his 1844 *Handbook of the Isle of Wight* provided a convincing explanation, when he wrote of Ryde's pier:

To the timid, and to those who are deterred by other causes than fear from venturing on the 'heaving wave', the Pier affords innumerable attractions. The arrival and departure of the steam packets;—the numerous boats everywhere sailing about;—the merchantmen constantly underway;—together with the occasional naval salutes, announcing the arrival or departure of ships of war, compose a scene of unusual interest and excitement. Nor is the spectacle on the Pier itself the least attractive object, from the number and often from the elegance and beauty of the fair promenaders. A more delightful scene can scarcely be conceived than this Pier affords when the placid brightness of a summer's moon rests upon it. The combination of motion and stillness which the sea presents on a fine and tranquil night is inexpressibly pleasing.

6 *left* In many cases,
piers found as much
use as promenades as
landing stages;
Brighton Chain Pier,
c. 1825

7 *below* The opening
of Southport Pier in
1860. This has many
claims to being the
first true pleasure pier

With the 1840s and 1850s came a change in the clientele of these resorts. The wealthy, with time and money to spare, continued to dominate the scene, but the middle classes were beginning to control more and more of the nation's wealth and power as Britain's industrial and commercial might became firmly established. They began to take advantage of their leisure time and seaside holidays became part of their lives.

It was the railways, which allowed them to do this. From small beginnings the companies had expanded to lay thousands of miles of track, many of which linked the cities and large towns to the coast. Trains grew yearly faster, safer and more comfortable. Such people as Thomas Cook hired whole trains for church societies and the like, in addition to the excursions offered by the railway companies themselves. A day or two, even a week by the seaside, was an attractive and feasible proposition for ever more people, although the majority of the working class were still tied to their local area by shortage of free time and money. With this increased patronage of the resorts, it was only a matter of time before a true pleasure pier was built.

WELCOME STRANGERS

SUCCESS TO THE PIER COMPANY

As with all 'firsts', there is a dispute as to which pier may lay claim to being the first true pleasure pier. In Great Yarmouth, suggestions to build a promenade and landing pier had been made as early as 1843, but there seemed to be little general enthusiasm at the time. The idea was revived nine years later when Charles Palmer proposed to erect a pier by public subscription, both as an attraction in the town and as a memorial to Wellington. Again, nothing was done but shortly afterwards the Great Yarmouth Wellington Pier Company was formed, shares were issued, and by June 1853 Parliamentary Assent had been obtained for the erection of a pier.

Work started on the structure later that month and the 623ft pier featured ornamental railings and a 100ft long promenading platform at its extremity. It was opened four months later. Admission fees were 1d. for adults, ½d. for children, and in its first year the pier made £581 7s. 1d. from the sale of tickets. Outgoings (including 17s. 6d. for the piermaster and 7s. 6d. for his assistant, but excluding directors' fees) amounted to £170. Competition soon came from the Britannia Pier, erected a short distance away in 1858, and the Wellington Pier Company, faced with increasing maintenance, repair and running costs and decreasing gate receipts, never recovered its financial health.

Meanwhile the owners of the Margate Landing Place were also having problems. In the late 1840s, the pier was suffering so much storm damage that repairs were costing over £300 per annum. On 4 November 1851, the structure was breached in two places, one gap being 70ft long, the other 34ft. The Company decided that the most sensible course was to demolish the jetty and to erect a new pier. Work started on the new jetty in 1853 and it was opened in April 1855, although oddments were not completed until 1857. The contractor, Bastow,

PIERS OF THE 1860s

8 *left* Worthing (shown here c. 1866) followed Southport in acquiring a pier in 1861–2. Like many of the period it was a simple, unadorned structure

9 *above* Slightly more sophisticated was Blackpool North Pier, opened with great ceremony in 1863

10 *right* Brass plate of the Weston-Super-Mare Pier Co.

11 *below* From the early days, storms and ships inflicted damage on piers: Great Yarmouth Britannia Pier suffered severely in 1868

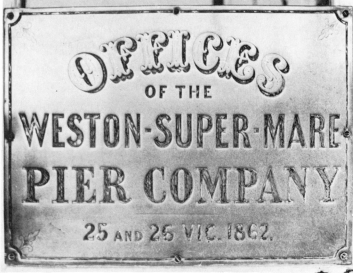

entered a bid of £10,750 (a fixed price contract) for the erection of the pier; the directors of the Company must have been astonished to find his the lowest tender by £12,000. His actual costs amounted to £15,400, but he was dismissed in 1856 for failing to make reasonable progress. One final item of expenditure was that of six guineas to the Coalbrookdale Company for casting the commemorative plaque.

The first of the pleasure piers

The completion of these structures marked the end of the embryonic stage of the development of piers. The Margate pier was of iron, and gave a pleasant and spacious promenade to the town. Nonetheless, it was a direct replacement for the earlier landing stage, and the strongest claim for the first pleasure pier probably lies on the other side of the country, at Southport.

The citizens of Southport had conceived the idea of a pier in the 1840s, although there were then two schools of thought concerning its purpose and use. The town had grown into a popular watering place and a pier was seen as an extension to the promenade to cater for the residents and the increasing number of visitors. The alternative plan was for a pier with railway to act as a commercial and trade scheme which would turn Southport into a port for the supply of goods to Manchester. Perhaps it was fortunate for the long-term development of the town that it was the former plan which was adopted. At a meeting in the Town Hall in 1859, a company was formed to build the pier, with a capital of £8,000. Work started in the same year, and the pier was completed and opened in 1860, the occasion being marked by a grand gala, procession, banquet, fireworks and a ball. It was an immediate success. Not only were there crowds of people willing and eager to pay their tolls for the chance of a stroll high above the beach and the waves, but steamboat passengers provided a respectable income. The pier was a regular port of call on the route which serviced Llandudno, Blackpool, Fleetwood and Lytham. Steam passenger vessels, notably the paddle steamers, were to become an essential element in the story of nearly every seaside pier.

Other resorts were not slow to grasp the advantages of having a pier. Worthing acquired one in 1861-2, a simple structure carrying no buildings but making an excellent promenade. At Blackpool, the North Pier was erected in 1862-3. It was built by the Blackpool Pier Company Limited, which had been formed in February 1862. This had its origins in a meeting in the Clifton Hotel, Tabot Square, in 1861, when leading townspeople had discussed the possibility of erecting a pier. The proponents' prospectus stated their object was to provide 'greater promenading space of the most invigorating kind', and to do this they wished to erect 'a substantial and safe means for visitors to walk over the sea to a distance of thirteen hundred and fifty feet at high water'. Twenty townsmen subscribed the preliminary expenses of forming a company which was started with a capital of £12,000 in 2,400 shares of £5 each. Thereafter things moved swiftly. Designs were called for and approved, a contract signed for the erection of the structure and the first pile placed in June 1862. The pier was opened just under a year later on 21 May 1863. The *Illustrated London News* of 30 May 1863 carried a full report of the proceedings. 'From an early hour in the morning', stated the report, 'the trains, both ordinary and special, discharged large numbers of visitors, and by noon, when the proceedings of the day may be said to have commenced, the number of strangers was little short of 20,000.'

The appearance of the town itself was as gay as any holiday could make it. Streamers, flags, and banners almost hid the thoroughfares, the shops were closed, and everyone appeared to join in the festivities. The pier and its immediate vicinity formed the great point of attraction. The ceremony of opening the pier began with a procession, which represented all the local interests of the town. The marching of the procession occupied about two hours and, after passing through the chief streets and round the pier, it halted opposite the Clifton Arms Hotel, where Mr. F. Preston, the chairman of the pier company, said that the pleasing duty devolved upon him, on behalf of the Directors of the Blackpool Pier Company, formally to open the undertaking which they commenced last year. The crowd gave several hearty cheers, and the demonstration was increased by a salvo or artillery. . . .

A dejeuner was afterwards served in the Clifton Hotel, to which about 150 persons sat down. Mr. Preston occupied the chair, and after a few speeches from him and other gentlemen, the proceedings terminated.

This scene was to be repeated, with minor variations, scores of times throughout the United Kingdom in the next forty or so years. The object of all the attention on this occasion was an elegant structure 1,070ft long, with an abutment and approach giving a further 200ft of promenading area. Several shelters and refreshment stalls were carried on projecting parts of the deck and comprehensive seating and lighting was provided for the convenience of the promenaders. Steamboat passengers were catered for by means of a series of landing stages connected with each other and with the pier by broad flights of steps. But even as Blackpool Pier was being opened, work was starting on similar structures elsewhere.

The West Pier at Brighton started to rise alongside the Chain Pier in 1863. Enthusiastic shareholders watched its erection over the next three years, and it was indeed a lavish pier that was opened in 1866, of similar length to Blackpool's. It not only had an enlarged deck at the seaward end, or head, but also at the landward end, to increase the deck area available for promenading. Ornamental gates and toll houses led on to the pier. The first section, built at a higher level, was connected to the rest by an incline for bath chairs, and further ornamental shelters were provided along the rest of the pier. The head featured a series of weather screens—another convenience for the promenaders, the paying customers.

Elsewhere the paying customers were flocking through the turnstiles. The pier at Deal was built in 1863–6, and that at Aberystwyth in 1864–5. Construction of this latter pier proceeded simultaneously with those at Bognor and Lytham. The philosophy behind each of these piers was basically the same. A pier was regarded partly as an investment, a business venture in itself, and partly as an attraction to draw visitors, and therefore trade, to the town. Aberystwyth, and the other pier towns of the mid-1860s, were growing resorts where a pier was considered not only an asset, but almost an essential element in their future development.

The ensuing few years saw the familiar pattern of public meetings, formation of companies, issuing of shares, letting of contracts and opening of the resultant piers in resorts up and down the country. Teignmouth (1865–7), Eastbourne (1866–72) and New Brighton (1866–7), were among their numbers. Scarborough was another resort taking its first steps into the pier age.

A fishing, trading and ship building port until the late seventeenth century, Scarborough had started to change with the discovery of the spa. The wealthy holidaymakers of the early 1800s were enticed to the resort by such facilities as the sea bathing infirmary of 1804. Development of the town was assisted by the formation of the Cliff Bridge Co. in 1826 which opened the Cliff Bridge the following year thus giving easier access to the spa. Tollgates were erected to keep 'improper persons' away, and so ensure selectness. Scarborough benefited (suffered, some might say) from Thomas Cook's railway excursions of the 1850s which brought trainloads of Midlanders to sample the sea air and the sands. But local notables felt that something was lacking and formed a company to build a pleasure pier. The apparent success of the established piers elsewhere must have been an extra incentive to the promoters of the scheme.

The pier was started in 1866 and opened some three years later at a cost of £15,000. Initially the pier was a financial success, and a small dividend was paid, but its subsequent history was not happy and its life was only some forty years.

Other piers of the 1860s fared slightly better. One unusual example was at Weston-super-Mare, opened in June 1867. This incorporated an island within its design. A promenade pier bridged the 1,100ft between the abutment and the islet, which was levelled, stepped and bricked to form a further promenade. A second section of pier, some 250ft long, struck out at right angles to act as a landing stage to snap up the steamboat traffic.

The structure was complemented by that at Clevedon, a few miles further up the coast. Here, as at Southport, the pier had its origins in a scheme to promote Clevedon as a major commercial port, this time on the London/South Wales route. In 1800 Clevedon was a rural community of about 300 persons, but the arrival of the railway in 1847 and its proximity to the major port and manufacturing city of Bristol ensured its growth as a holiday resort. The need to cater for its visitors and the desire to broaden the base of Clevedon's economy resulted in the formation of the Clevedon Pier Company. In the event it was a promenade with steamer landing stage that resulted. The original Act for the pier's construction was obtained in the parliamentary session of 1863–4, but work did not commence until late 1867; most of the building took place, under appalling conditions, in 1868. The whole town seemed to be *en fête* for the official opening in April 1869, and the British coast officially acquired yet another promenade pier.

Successes and failures

By the end of the decade, other piers were either open or under construction. Rhyl, Saltburn, Douglas, Hastings, Hunstanton and Morecambe joined the resorts able to include eloquent descriptions of piers in their guides and brochures.

In these towns the preliminary work of promoting a pier had been greatly eased by the passing of the General Pier and Harbour Act of 1861, and Parliamentary assent for pier construction was usually forthcoming, including permission to impose tolls. Southport's rates (1868) for example, included the following:

For every person who shall use the pier for the purpose of walking for exercise, pleasure, or any other purpose, except for embarking or disembarking, for each and every time .. .. 6d.

For every bath or sedan chair taken on the pier, for each and every time. 1s. 0d.

For every perambulator .. .. 6d.

For every person using the tramway for each single journey and exclusive of luggage .. .. 3d.

Rates on passengers' luggage

For every trunk, portmanteau, box, parcel, or other package within
the description of luggage, not exceeding 14lb. in weight   ..   ..   1d.

Over 14lb. and not exceeding 56lb. ..   ..   2d.

Over 56lb. and not exceeding 84lb. ..   ..   3d.

Over 84lb. and not exceeding 150lb.   ..   ..   4d.

And for every additional 50lb.   ..   ..   2d.

It was not all straightforward speculation, however, and in this respect the pier enthusiasts of Deal were particularly unfortunate. A company for erecting a pier at Deal had originally been formed in 1838, it being planned to erect a structure 415ft long. The approval of the Board of Admiralty was secured and an Act of Parliament obtained to raise £21,000 in 4,200 shares of £5 each. Work commenced on the pier, and 250ft was completed at a cost of £12,000. However, the company ran into financial trouble, work stopped, and the shareholders had the distressing experience of seeing the fruits of their investment gradually succumb to decay and storm, until in the late 1850s the tattered remains of the pier were washed ashore in a sudden squall. The materials were auctioned off and netted only £50.

Nonetheless, when proposals were again made to build a pier, there was sufficient interest to get the scheme under way. The result was a 1,000ft iron structure which the Deal and Walmer Pier Co. were unable to pay for, and possession passed to the builders in 1866.

Other setbacks were more dramatic. *The Times* of 4 and 5 August 1867 reported panic among the promenaders on Brighton's West Pier when the structure noticeably shuddered and oscillated several times. A correspondent wrote of ladies fainting and children being trodden underfoot as the crowd hurried to the pier entrance, fearing that the structure would collapse at any moment. The secretary of the pier company dismissed the affair as being within the design limits of the pier and of no concern, although a third observer suggested that a steamer using the pier as a 'brake' when mooring was the cause of the vibration. Certainly the story seems to have alarmed some would-be promenaders and receipts at the turnstile are reported to have dropped. But the setback was only temporary; the incident was soon forgotten and the pier regained its former popularity.

The owner of the Aberystwyth Pier may well have had some sympathy with the trials of the Brighton Pier promoters. The Welsh pier had been opened on Good Friday 1865, but in January of the following year a hundred foot of the structure was washed away in the storm. Repairs were not effected until 1872, when the structure was bought by the Aberystwyth Pier Company.

Pier owners and shareholders everywhere, however, must have been cheered by the increasing popularity of the seaside. Whilst it is true that holidays with or without pay were for many a long way off, the excursion trains continued to pour thousands into the resorts for the day out in addition to those who could spend more time there. The response of the pier companies was to increase the appeal of their structures by providing more facilities for their patrons. Many piers were equipped with refreshment booths and small stalls, and often a bandstand. New Brighton Pier had a 'handsome saloon' with observation tower. Bands were engaged to play on the pier heads, and the turnstiles clicked over in a most satisfactory manner.

Such was the success of some piers that extensions and alterations were made to meet changing or increasing needs. At Southport, there were three points of criticism, viz the lack of waiting and refreshment rooms, the difficulty experienced by ladies attempting to negotiate their voluminous dresses through the turnstiles, and the long walk along the 3,600ft pier that then faced them. The pier company did its best to oblige its patrons. Refreshment and waiting facilities were provided in 1862, and on 7 May 1863 a pier tramway was opened. This was by no means a new idea—the Herne Bay and Ryde piers featured this means of conveyance. In the Southport scheme, the vehicles were at first moved by hand. Complaints that the tramway was interfering with the use of the pier as a promenade prompted the company to widen the pier and to relay the tramlines on the side of the structure, fenced off from the rest of the deck. The opportunity was also taken to incorporate a stationary winding engine to propel the cars. These vehicles, open affairs with 'knifeboard' seating, took the passengers along the entire pier in three minutes and became a popular attraction.

Sadly, there was a fatal accident in August 1865, when one of the two cars left the track. Repairs and inspections were subsequently carried out and, as at Brighton, the confidence of the public was soon restored.

As for the third criticism (the difficulty in passing through the turnstiles), the pier company smugly suggested that ladies experiencing this problem should buy a season ticket which entitled them to use the gates.

At Blackpool, the North Pier was proving such a success that not only was it extended, but plans were made for a second pier. There were 130,000 visitors to the town in 1861, and this had increased to 800,000 during six months of 1868. To cater for them, the new pier (known at the time as the South Pier, but later as the Central Pier) was started in July 1867, and opened in May 1868. The main section of the pier measured 1,118ft and there was an additional 400ft low-water jetty at the seaward end. A similar feature was added to the North Pier in 1867.

By the end of the decade, the seaside piers were becoming established features of many resorts, and the resorts were becoming daily more popular. As *The Times* of 30 August 1860 commented:

> Our seaside towns have been turned inside out. So infallible and unchanging are the attractions of the ocean that it is enough for any place to stand on our shore. That one recommendation is sufficient. Down comes the excursion train with its thousands, some with a month's range others tethered to a six hours limit, but all rushing with one impulse to the water's edge.

The piers became a natural focal point in those towns that possessed them. Through the turnstiles lay a new world, away from the dust and heat of the town. The Briton's inherent passion for things nautical was satisfied by a typical British compromise. On the pier he (or she) could enjoy all the advantages of going to sea with none of the attendant danger or discomfort. He could stroll along the deck, peering through the boards to see the swirling water beneath. As he inhaled the salubrious air he could scan the horizon for a sail or smudge of smoke which would indicate a part of Britain's naval and mercantile might. He could gaze back at the shore and see the cliffs and beaches for miles in both directions, and possibly the landscape behind the town. He could do all this, and never have to worry about feeling queasy, about getting his feet wet, or about not getting back in time for tea. It was an experience unlike any other.

Postscript

A fitting postscript to this period is the following announcement from the Saltburn-by-the-Sea Pier Company:

Secretary's Office
Saltburn-by-the-Sea.

November 1869.

The Directors have arranged to exhibit advertisements in this delightful promenade at a low scale of charges.

The immense number of people who pass over the Pier both Visitors and Residents in Saltburn renders it one of the best possible stations for advertising and as a rule, Visitors will have more time to read and notice advertisements whilst inhaling the sea breeze and enjoying the glorious prospect of land and water scenery and the unequalled expanse of sandy beach than the ordinary Railway Traveller who is either intent upon catching the train, or keeping some appointment necessitating dispatch, and leaving neither time nor inclination for looking about, which is proverbially what people go to the Sea-side for.

I may mention that in something less than six months, namely during the time the Pier has been opened to the Public upwards of 50,000 people have paid to walk upon it and it is reasonable to expect that this number will be more than doubled during the next season.

Applications for space must be addressed to the Secretary as early as possible as allotment will be made in the order in which application is received. The lowest charge will be 5/- for the year (or any portion of the year) between the 1st January and 31st December 1870. Auctioneers, Theatrical Managers and others can arrange to have their bills posted periodically without personal attendance.

By Order
Thos. C. Tomkins, Sec.

TWO

THE PIER BUILDING ERA

Any account of the development of seaside pleasure piers in the last thirty years of the nineteenth century must necessarily be selective. It was a period when about fifty piers were built, when others disappeared and when many were extended, improved or otherwise altered. Some piers are particularly worthy of mention whilst others are unremarkable.

This period was also one in which seaside resorts underwent drastic changes. This was due in no small part to the provisions of the 1871 Bank Holiday Act. It was not radical in itself, but it had far-reaching effects. Basically, the Act guaranteed a limited number of workers the right to a stated number of holidays, three of which had been widely observed for some time. The roots of this move were in the Factory Act of 1850, which prohibited the employment of women and young persons in a factory after 2 p.m. on a Saturday, and also in the summer closure of some factories. This latter practice, widespread in the industrial North-West, enabled machines to be overhauled and gave the factory workers an annual break from their labours. The culmination of these three factors was the growing acceptance of annual holidays throughout British industry and eventually holidays with pay.

The more immediate effect was to present the seaside with a new phenomenon—the August Bank Holiday. The catchment areas of the resorts were yearly expanding, both geographically and socially. Trainloads of visitors and day trippers poured into the resorts on this more than any other day. The pretence of visiting the seaside for health reasons was replaced by the honest assessment of a day by the sea being for pleasure and for fun. The new generation of holidaymakers who flocked to the seaside in the 1870s, '80s and '90s were not content to stroll along the promenade, to contemplate the motions of the sea and to make idle chit-chat with their companions. Certainly, such organisations as the National Sunday League and temperance movement encouraged and sponsored excursions to improve the moral condition of the workers, to remove the temptations of beer shops and to foster the right environment for the development of those Victorian ideals of cleanliness, discipline, and self-improvement. But for many people, the seaside was the one place where they could escape from the stifling confines of Victorian urban life, where they could enjoy themselves without fear of reproach or reprimand and where the monotony and frustrations of fifty-odd weeks of boredom could be fully

dissipated. The need was for entertainment.

The resorts, for the most part, quickly responded to the changing pattern of public demand. Piers were in the forefront of this response. They were constructed and altered to provide the entertainment required, the organisations behind them were changed to suit and the piers' functions as promenades and landing stages slipped into a secondary role. Instant success for all piers, however, was by no means assured by this trend, and there were failures and disappointments as well as triumphs. The piers of two counties, in fact, gave a fairly good illustration of these extremes.

The Piers of Sussex and Yorkshire

At the start of the 1870s, Sussex had five piers in use (one of which was open although still under construction) with a further pier in the course of erection. The Brighton Chain pier of 1822–3, although not a financial success, was nonetheless a popular attraction in the resort. Refreshment rooms, shops and a reading

room, as well as a camera obscura, had appeared in the 1830s and 1840s to cater for the promenaders. The piers at Bognor and Worthing were little more than promenades, although those at Brighton (West) and Eastbourne featured ornamental kiosks housing refreshment and souvenir stalls, with such elementary fixtures as weatherscreens and continuous seating.

Hastings Pier, opened in 1872, was of similar design but incorporated a pier-head pavilion with a capacity of 2,000 persons. No longer was pier entertainment limited to a band playing in a small bandstand exposed to wind and rain. Concerts, musicals and plays could be performed in comfort.

The success of this pier had its effect on promoters of other Sussex piers. In 1887–9, Worthing Pier was greatly extended and a pier-head pavilion added, and Eastbourne acquired a pavilion in 1888. Extensions were made to the Brighton West Pier in 1893, and a pavilion opened there. Six years later work commenced on a second pier theatre and two games saloons at Eastbourne Pier.

The pier enthusiasts of Sussex not only improved existing piers, but constructed two new ones. The first was St Leonards, built in 1888–91 at a cost of

£30,000. The pier, 960ft long, featured a 700-seat pavilion built at the shoreward end and approached from the shore by a carriage drive. The pier was hailed as a masterpiece of design and construction, and certainly was a contrast with the bare structures erected some thirty years previously.

In the 1890s the Chain Pier at Brighton was giving cause for concern. Seventy years' exposure to wind and sea had taken their toll, and the pier was both a financial and a structural liability. In 1889 the pier was bought by the Marine Palace and Pier Co. for £15,000, the shareholders in the original company receiving for each £100 share £13 6s. 8d. in cash and £36 13s. 4d. in debentures of the new company. This latter organisation proposed to construct a new pier at Brighton and Government consent to do so was conditional upon the Chain Pier being demolished. The old pier was closed to the public in October 1896, at which time the head was 6ft 9in out of plumb. The pier did not long survive. A powerful storm in December 1896 destroyed the Chain Pier as well as causing serious damage to the shore end of the West Pier. The demolition teams dismantled what little was left of the structure, and an auction of the wreckage raised a few pounds for the owners.

The structure which was rising further down the beach to replace the Chain Pier was a complete contrast in every way. Designed from the beginning as an amusement and pleasure emporium, the Palace Pier was 1,700ft in length (500ft longer than the West Pier) with a wide promenade deck and a pier-head pavilion accommodating 1,500 people. Dining and grill rooms, with smoking and reading rooms, complemented the pavilion. There was provision for bathers beneath the pier head and landing stage for pleasure craft completed the structure. It was, as a contemporary report stated, 'unequalled by any similar undertaking in the United Kingdom'.

There was one other pier in Sussex. This was at Bexhill, where work started in 1898. However, only a small part of the pier was built and difficulties led to the project being abandoned shortly afterwards.

In the North, the high degree of success of the Sussex piers was being equalled by the misfortunes and setbacks suffered by the Yorkshire piers.

The Hornsea Pier Company had been formed in 1865, but there was a considerable delay in starting work on the structure. In 1879, the pier was under con-

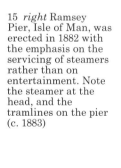

15 *right* Ramsey Pier, Isle of Man, was erected in 1882 with the emphasis on the servicing of steamers rather than on entertainment. Note the steamer at the head, and the tramlines on the pier (c. 1883)

16 A different structure was erected at Sea View, Isle of Wight (1880), shown c. 1900. One of only two suspension piers built, it survived until the 1950s

struction but in the following year a small ship collided with it, and 200ft of ironwork was carried away, in addition to which the pavilion at the seaward end was destroyed. Protracted litigation followed whilst the pier continued to be battered by heavy seas. It was reported derelict in 1897 and the pier finally disappeared in 1910 when the promenade was reconstructed.

Scarborough's pier, following the ship collision in 1883, was sold to a London businessman for £1,240. He spent £10,000 on repairs and improvements, including a refreshment bar and the pierhead pavilion which attracted famous artistes. Despite this, receipts were disappointing. The attractions of the well-established spa area of the town and the fact that there was no made up road giving easy access to the pier may have been responsible. In 1904 the pier changed hands again. But like the Chain Pier in 1896, Scarborough Pier was washed away overnight in January 1905, and since the loss was not covered by insurance the pier was never replaced.

Saltburn was another Yorkshire town which had realised the desirability of a pier. It was financed by the Saltburn Improvements Company and erected in 1868–70; with a length of 1,400ft, the landing stage and part of the pier itself were destroyed by storms in 1874 and 1875, and by 1880 the pier company offered its one time pride and joy for sale by auction. As in the case of Scarborough, the new owners attempted to revitalise the structure and restored it to a length of 1,250ft. Windscreens, a small theatre, a bandstand and refreshment and retiring rooms were added, the Cliff Lift was restored to give easier access to the pier, and by the turn of the century it was a popular feature of the resort.

Meanwhile, the twin towns of Redcar and Coatham had acquired piers. The Redcar 'Pier Company had been formed in 1866 for the purpose of providing 'a commodious promenade and landing pier', and Parliamentary assent obtained. Nothing was done actually to build the pier, however, until a rival scheme was proposed for nearby Coatham in 1870. There were attempts to reach a compromise and build a pier which would serve both resorts, but the two bodies of proponents could not agree on a site, and the construction of two structures went ahead.

Coatham Pier was started in 1873 and whilst still under construction it was badly damaged in November 1874 when the brig *Griffen* was driven through it in a storm. A second breach was made by the *Corymbus* in the same storm. The

17 Southend photographed c. 1910. The new pier of 1887, which included the clock tower, was extended in 1898 when a new head was built. In 1908 the upper deck was added

proposed 2,000ft pier was eventually completed in 1875 with an overall length of 1,800ft, with two pavilions. It did not last long. In 1898, the Finnish barque *Birger* was driven through the pier in a storm and the Coatham Pier Company, which was already in difficulties following earlier repairs, demolished the isolated section. The following year, the company crashed, and their pier was dismantled for scrap.

Redcar Pier was a less ambitious but slightly more successful structure. The structure was built in 1871–3 and its overall length of 1,300ft was adorned only by a toll office, ladies' and gentlemen's rooms and a shop at the entrance and seating and screens on the pier itself. However, in October 1880, the brig *Luna* cut the pier in two and damage was estimated at £1,000 (the cost of erection of the pier being only £6,250). In 1885 the landing stage was demolished by a steamer and the pier company was unable to finance repairs. Further damage occurred in 1897 when wreckage from the schooner *Amarant* collided with the pier. The pier head was badly damaged by fire the following year. Nonetheless, despite these misfortunes, the pier was in a usable condition at the end of the century, unlike its near neighbour.

29

The last of the Yorkshire piers was at Withernsea. This pier was completed in 1875 and the pier owners, the Withernsea Improvement Company, did not have long to wait before the same fate befell their structure as had befallen the others. Two hundred feet of the 1,200ft structure was carried away in the storm of October 1880 which had severed Redcar Pier. Repairs were effected, but further damage occurred in 1882. By 1890, only 300ft of the pier was left in a usable condition, and despite extensive reconstruction only 50ft remained in 1900. Subsequent repair work to the sea wall soon left only the ornamental gates as a reminder of another ill-fated venture.

Pier entertainments

Pier operators were quick to grasp the potential of the attractions of live entertainment shows. Some comparatively early piers incorporated pavilions where concerts, plays and musicals could be held. Bands were complemented by orchestras. At Eastbourne, Mr Wolfe's German Band was engaged to play in the pier's first season. The company of twelve received £3 for performing four times a day, seven days a week. This was soon reduced to a six day week when local opinion led by the Duke of Devonshire (the chief shareholder in the Pier Company) objected to Sunday performances. When Mr Wolfe moved on, the sixteen-man Hanoverian Band was commissioned to perform twice daily for the entire season (1 June–31 October) for a £250 fee.

At Blackpool, the North Pier acquired its first pavilion in 1874 under the name of the 'Indian Pavilion'. Its concerts became famous throughout the country, and attracted the attention of eminent critics. Orchestral music of the highest order

followed upon the appointment of Edward de Jong as conductor of the pier orchestra. De Jong, the most eminent flautist of his day, was followed by Risegari, and he was succeeded in 1883 by Simon Speelman. Speelman, leader since the opening of the pavilion, was later known as an eminent viola player of the Manchester concerts, the Halle Orchestra and the Brodsky Quartet.

These were but a few of the orchestras which performed at piers throughout the country. For those with other tastes, however, alternatives were provided. There were artistes whose versatility extended to singing, dancing, reciting monologues and giving 'character readings' from the classics.

There was comic opera. Gilbert and Sullivan's *Pinafore*, under the title of *H.M.S. Pinafore on the Water*, was played at Paignton Pier by Mr D'Oyley Carte's full company on 27 and 28 July, 1880.

There was also music hall with such famous performers as Will Catlin, who appeared at Scarborough Pier, and Dan Leno, Lily Langtry, Mary Randall and Arthur Roberts, who were on the bill at Folkestone Pier.

Pierrots were a further important feature of pier entertainment. They were groups of entertainers, in fancy dress, with a varied repertoire. They grew up as successors to the black minstrels and often worked the beaches and streets of seaside resorts, as well as the piers.

Visitors thrilled to the daring men who dived off the piers. Nearly all bestowed upon themselves the title of 'Professor' and Professors Capes and Connell entertained the crowds at Skegness in the 1880s and '90s. At Rhyl in 1887, Tommy Burns, the World Champion diver, was engaged. He was a famous and obviously talented man. Besides his string of world records for swimming and diving, Burns had saved a remarkable number of people from drowning, and about the whole town turned out for his arrival on 16 July for the opening of his season. His journey to the pier was via several public houses and he was obviously indisposed when he arrived there. He had, apparently, some difficulty in keeping his feet, but, being the showman he was, he insisted on giving his diving performance. He launched himself from the end of the pier, and was instantly killed.

Less dramatic but equally popular were the dances held in the pavilions and on the open decks. A warm clear night on the pier provided the perfect romantic setting for the young Victorian ladies and their swains, and it was as near as many would get to a luxury cruise liner.

As an example of a typical pier show, the week commencing 18 August 1890, at Skegness saw the following series of acts for the entertainment of pier patrons there:

> Captain Slingsly, ventriloquist
> Walter Howard, banjoist
> Signor Isidore Corelli, in 'Silent Sorcery'
> and Miss Myfanwy Morris, contralto

as well as minstrels and divers.

As in many places, the 'Sacred Concerts' on Sunday were free, although tickets for the more secular entertainment were 1s. and 6d. (including pier admission).

There were moments of light relief. The promenade pier at Ramsgate had a camera obscura in the head, and one of the customers in the nineties was a London C.I.D. man on holiday. Whilst viewing the device, he saw in the 'living panorama' of the sands a well-known crook from town, busily engaged in picking pockets. An arrest and subsequent conviction followed.

The paddle steamers

Perhaps the one feature of pier amusement which has made the greatest impression on history was the paddle steamer excursion. The coming of the railway to almost all resorts by the 1870s drew much of the traffic from the steamboats as a means of reaching these places. However, the vessels found an equally successful use in the provision of excursion trips. From a jaunt across the bay to a day trip to France, the paddle steamer provided the climax to many a day on the pier.

At Skegness, steam boat trips started in 1882 shortly after the pier was opened. The following year, the Skegness Steamboat Company was formed by local businessmen, to provide trips around the Wash. The landing stages around the pier head permitted the embarkation of passengers for three hours either side of high water, and by 1883 the paddle steamer *May*, one of the largest on the east coast, with a capacity of 255 persons, was on station. Excepting Sundays and barring bad weather, this vessel and others provided a variety of trips for holiday-makers. A one hour cruise along the coast cost 1s., whilst for a further 6d. passengers could take an alternative $2\frac{1}{2}$-hour trip to the Lynn Well lightship, which included a tour of this vessel. These excursions had the secondary function of supplying the lightship crew with newspapers and fresh vegetables.

Undoubtedly the most popular trips of all were those to Hunstanton on the Norfolk coast across the Wash. These left Skegness at about 8.30 a.m. and arrived at the Norfolk resort at 11.00 a.m. Most of the passengers then took the opportunity to visit the royal residence at Sandringham; some took the train to Wolfreton, whilst others formed themselves into groups and hired carriages for the eight mile drive. The return journey deposited the travellers back at Skegness at 8 o'clock at night. The cost was 3s., plus pier fees at Hunstanton.

Skegness was one instance where many visitors found it more convenient to travel by sea than by rail. Paddle steamers with such names as *Scarborough*, *Yorkshire Lass*, *Buster*, *Spindrift* and *Privateer* brought groups of trippers from the Wash ports in the morning and returned them at night, as well as taking others on their trips to sea.

Paddle steamers were associated with nearly all the other piers. Services connected those on the Sussex and Kent coasts, for example, and trips to the Continent

21 Eastbourne Pier in 1895. The pierhead theatre dated from 1888, and later additions were made in 1899 and 1901

and the Channel Islands were amongst those offered to the more discerning and adventurous of holidaymakers. The inspiration behind Britain's naval and mercantile powers was reflected in thousands of paddle steamer excursionists.

The pier mania

There were, of course, other piers. A pier building mania gripped seaside Britain. Piers were built almost annually. There was an unmistakeable rivalry between resorts to build bigger, better, more sophisticated (and, it was hoped, more profitable) piers, which would bring in more holiday trade. The appeal of Victorian technology, social need and British entrepreneurism seldom failed to raise the cash required to erect a pier. Public money was sometimes spent on piers if, for one reason or another, private enterprise did not sponsor a scheme. Some piers were erected as investments both in themselves and in connection with the trade it was hoped would follow.

Herne Bay's first pier, for example, had finally succumbed to decay and local indifference in the early 1860s. It had shared a chicken and egg relationship with the town, the latter developing in accordance with the traffic figures.

The want of a pier was felt, however, and the Herne Bay Promenade Pier Co. financed a modest structure (its length was only 320ft) which opened in 1873. The cost was a mere £2,000 compared with the £50,000 total for the first pier. It was designed purely as a promenade and pleasure pier, unlike its predecessor, and was improved by the construction of a pavilion with theatre and lock-up shops across the entrance in 1884. Admission charge at this time was 1d., a day ticket was 2d., whilst residents and long term visitors could purchase a season ticket for 5s. Experimental visits were made by shallow draught steamers in 1892 to gauge public interest in a deep sea pier, and shortly afterwards plans were drawn up for just such a structure. This pier, with a total length of 3,787ft, incorporated the 1873 pier and was constructed in the period 1896–9. At first the pier's finances were promising. The annual meeting of the shareholders of the Herne Bay Pier Co. in 1900 was told that pier tolls had been £1,758, refreshment receipts £1,254 and tram and pavilion income respectively £480 and £318. But the financial failure of the pier manager led to the liquidation of the pier company and the pier passed to the builders in the same way that others had done through the mis-management of their owners. Head Wrightson, the contractors in question, sent staff from their London office to man the turnstiles; the local council purchased the pier in 1909.

At Bournemouth, however, it was the local authority which had promoted the piers from the start. In 1851, the population of the district was only 695 but its rapid growth as a select watering place led to the establishment of a Board of Commissioners to administer the village under the terms of the Bournemouth Improvement Act of 1856. One of their first acts was (in 1856) to construct a small wooden jetty to facilitate passenger landings. Decay and storm damage led to its demolition in 1861. It was replaced by a more ambitious structure 1,000ft long with a T-head at the seaward end. It cost £3,418. One item of expendititure was the wage of the toll collector, Thomas Burt, which was 12s. weekly; he also had to be pro-vided with two suits of clothes and two hats annually, and a great coat every other year. The Commissioners aimed to recover their costs, and then to make a profit by charging admission tolls. Annual ticket charges were based on rateable value: ratepayers and their families rated at less than £15 p.a. paid 5s. for a yearly ticket; those with a rateable value of £50 paid 10s. 6d. whilst the ratepayer and his family (which included servants and near relations living in) of over £50 had to find £1 1s. 0d. Non-ratepayers could admit themselves and their families (not ex-ceeding six persons) for 10s. 6d. monthly, £1 1s. quarterly or £2 2s. annually. Unfortunately, as quickly as they made money, the Commissioners were spending it on repairs. Serious damage was caused by a storm in January 1867 and further damage in 1876 left the pier unserviceable. One item of expenditure in the follow-year was £17 19s. 11d. to Mr W. Hill for the recovery of wreckage of the pier which apparently had been washed ashore in Swanage! A temporary jetty was erected in 1877 to service the steam boats, whilst the townspeople decided what steps to take next.

The repeated mishaps, the constant expenditure and the other matters demand-ing the attention of the Commissioners did not encourage further commitments to a pier. A private company had, in fact, proposed the construction of a pier in the mid-1870s but negotiation between this organisation and the local authority came to nothing, and the construction of a pier went ahead at public expense. This was opened in 1880, and along its 1,000ft length there were alcoves offering shelter from the wind, with shelters and a bandstand at the far end. *Brights Guide to Bournemouth* of 1897 had the following comments to offer.

The Pier offers an excellent promenade and resting place for those who, while wishing to enjoy the sea air, do not care to run the risk of an attack of mal de mer. Those who have no fear of this unenjoyable state flock to the landing stages, whence the splendid steamers of the Bournemouth, Swanage and Weymouth, and Bournemouth and South Coast Steam Packets, Limited, all through the summer season, take hundreds of visitors to spend a few hours at one or other of the Watering Places on our Coast between Brighton or Torquay, or run shorter trips, at intervals during the day, for those who wish to explore the beauties of Swanage, Lulworth, Alum Bay, &C.

During the early hours of the morning numbers of bathers are attracted to the Pier, from the end of which they are enabled to enjoy the luxury of a dive into clear, deep water, from the springboard which is fastened at the landing stage; while in the evening, those who love to see the mantle of the night as it gradually clothes the earth can here watch the last rays of the sun disappearing behind the Dorsetshire hills, and catch a final glimpse at the twilight-shaded bay before returning to their homes.

The charge for admission, the guide then added less romantically, was 2d.

Another guide of the period commented favourably on the band performances at the pier head (twice daily morning and evening in summer, morning and afternoon in winter) but bemoaned the fact that there was no pier pavilion. The shelters were, it was pointed out, hopelessly inadequate to afford protection to even a moderate number of people in the event of rain 'which will occasionally fall even in the best regulated pleasure resorts'.

The town was not receptive to the idea of a pavilion. Contemporary reports alleged that generally 'dances and light concerts were discouraged; and dissipation is said to take the shape of bazaars and social meetings for charitable objects.' 'This most decorous and dull of watering-places' was another description of the town.

There were two other piers in the town. That at Boscombe was erected in 1888–9 as a promenade and steamer landing place. It remained privately owned until 1904, when it was bought by Bournemouth Corporation. Contemporary with this structure was the similar Southbourne pier. It was opened on 2 August 1888, the occasion being marked by the visit of the pleasure steamer *Lord Elgin* from Bournemouth pier. The programme of sailings between Southbourne and Bournemouth piers commenced four days later, with the *Lord Elgin* providing a daily service leaving at 9.30 a.m. and returning nine hours later. Return fares were 1s. first class and 9d. second class, although the two excursions on bank holidays cost only 6d. Successful though these trips were, they did not last long. Serious storm damage in 1900 and 1901 left the pier derelict. The Corporation declined the offer of the pier from the Southbourne Pier Company, and what was left of the pier was demolished as a dangerous structure in 1907.

Bournemouth might have been decorous and dull in the 1890s but Blackpool certainly was not. In the 1890s, Blackpool welcomed over three million visitors a year and pier accommodation was on a correspondingly grand scale. The North Pier, greatly extended in the 1860s, was further extended in 1875 and 1877 with the addition of north and south wings on the pier head; this area of this part then measured $1\frac{1}{2}$ acres. Electric lighting followed and widening of the pier in 1896–7 practically amounted to it being doubled in size.

The Central Pier was a favourite spot for enthusiasts of dancing and an 8,000 sq ft area for this purpose had been provided at the pier head. A central platform of the same type (measuring 380ft by 153ft) was added in the nineties.

The number of visitors, however, justified a third pier, and the Victoria Pier (later known as the South Pier) was erected in 1892–3. It was opened on Good Friday 1893 and admission figures were encouraging. 12,000 passed through the turnstiles on the opening day and figures for the following three days were equally impressive:

Saturday	11,000
Easter Sunday	5,000
Monday	13,000

Blackpool with its three piers was the ultimate in nineteenth-century pleasure

22 A more rudimentary affair was Mumbles Pier, shown c. 1905, built to cater for industrial Wales in 1897–8

resorts. These structures, with the tower, circus and winter gardens, all overlooking the famous golden sands, were living proof of the use of technology and engineering excellence to improve the quality of life for the Britons responsible for making Britain the workshop of the world.

On the other side of the country, two projects of interest were under way in the nineties. At Southend the saga of their pier had continued to 1846 when, after some false starts, an all-timber structure was finally completed to deep water, a total distance of a mile and a quarter. The pier passed to the Southend Local Board, who ran it at a profit for some years. In its fifty years of existence, the pier (in its various forms) had seen Southend grow from a village of a few hundred souls to a popular resort with a resident population of 7,000. Increasing holiday trade pointed to the desirability of a new pier. Parliamentary consent was obtained in 1887 and the new structure opened in August 1889. The only features of the old pier to be retained were the entrance booths, erected in 1885. The cost of £68,920 was increased by £21,000 in 1898 when an extension and new head were completed. The longest pier in Britain, and probably the longest in the world, measured up to a mile and a third.

Farther up the coast Clacton was engaged at this period in extending its pier. The original construction of the pier had been one of the first steps in the transformation of a sleepy fishing village into a holiday resort. Peter Schuyler Bruff, engineer and manager of the Eastern Union Railway, had purchased a large area of cliff land in the village and proposed to improve access to the area as a prerequisite to its development. Despite intense local opposition, a wooden pier (erected under Bruff's superintendence) was opened in July 1877, eleven years before the branch railway line was completed. The authorised toll of 2d. hinted at a promenading purpose, but in fact the pier was built for the landing of goods and passengers. Rates which would be levied included 6d. for a barrel of gunpowder, 1d. per cubic foot for musical instruments and £1 for a corpse.

The increasing popularity of the resort reflected the ease of access by rail from Liverpool Street Station and by the paddle steamers such as *London Belle* and *Clacton Belle* from London Bridge. The pier was inadequate to cater for this increased traffic and in 1890–3 the pier was lengthened to 1,180ft and incorporated a polygonal pier head complete with pavilion, with concert hall, stalls and refresh-

23 Margate Pier, c. 1900. An octagonal head and pavilion was added to the pier in 1875–8, and further additions in 1893 and 1900 reflected its popularity

ment and waiting rooms. A comprehensive landing stage was provided, and Clacton pier was the equal of any on the east coast.

The list of pier towns of England and Wales at this time read like a gazetteer of popular resorts. Besides those towns already mentioned, Cleethorpes, Aldeburgh and Walton-on-the-Naze joined the pier resorts on the east coast. Cleethorpes acquired its pier in 1875 when the Cleethorpes Promenade Pier Company erected it at a cost of £10,000. It was leased to the Manchester, Sheffield and Lincolnshire Railway Company in 1884 at an annual rent of £450; it was sold to this concern twenty years later for £11,250, as its trains brought excursionists to sample the delights of the pier and its Concert Hall at the seaward end. Walton-on-the-Naze Pier was built in 1895 to replace the 1830 structure.

Further round the coast, Ramsgate, Dover, Sea View, Sandown, Shanklin, Ventnor, Lee-on-Solent, Swanage, Paignton, Plymouth, Westward Ho! and Penarth entered the pier age. There was an unmistakable rivalry between resorts, with pier speculators attempting to ensure that their resort had the best pier in the area, and consequently, they hoped, the largest number of visitors. In the North-West, Llandudno, Ramsey, St Annes, Blackpool (South), Morecambe (West End) and Bangor were the sites of new piers. That at Rhos-on-Sea, erected in 1896, was, in fact, the same pier which had been erected at Douglas, Isle of Man, in 1869. It was dismantled in 1895, sold to a group of businessmen in the North Wales resort and erected just across the bay from where the Colwyn Bay Pier was opened four years later.

A visitor to any part of the coast was never far from a seaside pier. A writer in *Cassells Magazine* in the 1890s had this to say on the subject:

> ... if the population and individual wealth continue to increase as they have done of late years, while the means of locomotion go on improving, it will be necessary to alter the map of England, and represent it as a huge creature of the porcupine type, with gigantic piers instead of quills.

Misfortunes and mishaps

There were, unfortunately, sadder tales to tell. At Margate the Jetty was extended in 1875–7, the scheme including an octagonal head with pavilion. On 24 November 1877, a storm-driven wreck sliced through the pier, marooning between forty and fifty people at the seaward end. It was not possible to rescue them until the following day, and they spent an unpleasant night on the pier head. Damage was estimated at £4,000, and a claim of £7,000 was made against the owners of the wrecked vessel on the grounds of their negligence in not having the wreck removed. The case was tried before the Court of Queen's Bench in February 1879 but the claim was disallowed.

There was a more serious incident at Morecambe on 9 September 1895. Morecambe, linked by rail to the manufacturing towns of Lancashire and Yorkshire, had developed into a resort with two piers, erected in 1870 and 1893. The *Illustrated London News* of 14 September 1895 carried the following report of the accident:

> Here on Monday, September 9th, at eleven in the morning, the landing stage at the end of the pleasure pier, which projects far into the sea, was crowded with people, waiting to get on board the steam-boat *Express*, for an excursion to Blackpool. Part of the floor of this structure, composed of iron gratings sup-

ported by iron piers too slender for the unusual weight of such a throng, suddenly broke down beneath them; about fifty men, women and children were thrown into the water. It was not deep enough, on all sides, to drown them immediately, and many of them clung to the undamaged parts of the landing stage, or to the pier, until they could be relieved, there being no high waves. But the fall or shock had probably stunned a few of the weaker, and others had suffered contusions of the limbs, which made them unable to stand, while some endeavouring to reach the steam-boat, got into deep water. An elderly lady, Mrs. Ralph of Carlisle, was taken up drowned, and laid upon the deck of the steam-boat. Several other women, unconscious and almost lifeless when they were lifted out of the water, presently revived; but Clara Illingworth, wife of the caretaker of the Methley Board School, near Leeds, and a young man named Priestley, were drowned, their bodies not being recovered until low tide in the evening. Fractured legs and severe lacerations were suffered by three or four ladies, and there was one case of concussion of the spine, besides many injuries from the effects of the shock, or of the immersion, which might prove more or less serious. Yet the disaster might easily have caused a much greater loss of life.

Some notable ship strikes have already been mentioned. Fire was another hazard which took its toll of nineteenth-century piers. Southport Pier pavilion and other buildings were destroyed by fire on 18 September 1897, when damage to the tune of £4,000 was caused. (A new pavilion and other improvements were completed by 1902.)

In August 1898, the pier head at Redcar was damaged by fire. Pier ownership may have been both profitable and satisfying. However, to use nautical metaphors beloved by pier operators and their publicity agents, it was not all plain sailing.

Pier railways

There is one further aspect of pier operation which is of interest—the development of railways and tramways. Some were originally contractors' lines, used in the erection of the piers and were later adapted to carry freight and passengers' baggage to the pier head. The passengers themselves also often rode on the trains. In other cases, tramways or railways were provided as passenger lines from the start. Methods of propulsion varied. Early vehicles were usually hand propelled or horse drawn, although some (appropriately enough for waggons travelling out to sea) were fitted with sails! Steam engines, electric third rails, batteries and cables powered different trams at various times.

The railways on Southport and the first Herne Bay Piers have already been mentioned. The latter system died with the pier, and was not revived on the second structure. However, the third pier at this Kent resort incorporated a contractor's railway which was later utilised for passenger carriage and opened for this purpose in 1899. Ramsey and Southend-on-Sea Piers both incorporated railways, as did the new Walton-on-the-Naze Pier in 1898 and those at Rhyl and Bangor (although these last two did not long survive). The outstanding example, however, was at Ryde, Isle of Wight. A horse-drawn tramway started operation along the pier in 1864, and at various times thereafter steam traction was tried with little success. The line was linked to the Island railway system in 1871, but transhipment of passengers at Ryde Station was an inherent disadvantage of the scheme.

24 A severe storm in
1896 destroyed
Brighton Chain Pier
and badly damaged
the West Pier,
shown here

In 1877–80, therefore, a new railway pier was constructed alongside the old structure, and shortly afterwards electric traction replaced the horses.

Only a handful of piers incorporated railways, but they became a useful and entertaining feature of these structures.

As the century drew to a close piers were at the peak of their existence. As examples of an age and its social and technological outlook they rivalled the medieval churches and the country houses of the previous century. Despite the setbacks and problems, the pier owners and their patrons seemed well satisfied with the progress made during the pier era.

In 1898 it was reported to a sub-committee of Great Yarmouth Corporation (who were considering the erection of a pier) that many such structures were making handsome profits. The Margate Pier and Harbour Company had, in the previous year, paid over 6% interest on their stock and carried forward a balance of £1,689. The net profit of the Hastings Pier Company in that period was

£3,612 3s. 9d. and an 18% dividend was paid; Llandudno's dividend was $12\frac{1}{2}\%$, that of Southport $5\frac{1}{2}\%$, and that of the West Pier, Brighton, $7\frac{1}{2}\%$. The three pier companies at Blackpool paid dividends of 10, 8 and 4% with £100 shares in the North Pier Company worth between £275 and £278, the £10 shares in the Central Pier between £16 and £16 10s. and the £1 shares of the Victoria (South) Pier 19s.–20s. Clacton's dividend was 8%, compared to 4–5% prior to the erection of the pavilion. On the other hand, total balance of the Folkestone Pier and Lift Co. was £28 6s. $11\frac{1}{2}$d., the first profit for some years.

So successful, it seemed, were the seaside piers, that they had been copied on the other side of the Channel. Trouville, on the North French coast, acquired a promenade and landing pier in 1889, with the necessary pavilion. Blankenberghe, Belgium, opened a pier in 1895, which, from its appearance, could easily have been cast ashore after being cut adrift from Brighton or Blackpool or any other British resort. Imitation, then as now, was the sincerest form of flattery.

THREE

PIERS IN THE TWENTIETH CENTURY

Pier building continued into the early years of the twentieth century, and one of the first piers to be opened was at Cromer. Until the early 1890s the Norfolk town had been a quiet, rather select, resort (it had been connected by rail to London in 1877 and to the Midlands in 1887). Then land on the west front was put up for sale, the speculators moved in, and streets of hotels, boarding houses, shops and cafes were erected to lure the middle class holidaymakers to the town. Under the 1899 Cromer Protection Act, the Cromer Protection Commissioners (a body separate from the local council) became responsible for erection of a pier, and borrowed money for this purpose from the Alliance Assurance Company at 4% over sixty years. Work started in 1900, and the pier was opened on 8 June 1901 at a cost of £17,067 14s. 5d. At the end of the pier there was a small bandstand which in 1905 was extended to form a pavilion. Three years later its floor was covered with maple wood for roller skating, and the pavilion was later used for concerts and recitals.

Elsewhere on the same coast the pier age lived on. At Felixstowe and Southwold, the Coast Development Co. Ltd. financed the construction of piers in 1900 and 1905 respectively. This was speculative development designed both to enhance the amenities of these resorts and also to provide landing stages for the Belle steamers on the run linking the towns with such places as Yarmouth, Lowestoft, Walton, Clacton, Southend, Margate and London. In this respect these piers owed as much to the early pre 1850 structures as to the later pleasure piers for their existence, although both incorporated pavilions and refreshment rooms. And at Felixstowe, electric trains carried passengers along the half mile pier for 1d.

The same company also erected the Claremont Pier, Lowestoft, in 1902–3. This 600ft long structure received its first steamer, the *Walton Belle*, in May 1903, and Lowestoft entered the ranks of the day tripper resorts, a position enhanced by the completion of the Lowestoft–Yarmouth railway line in July. The T-shaped structure was shortly afterwards extended, and a pavilion added.

A few miles away Yarmouth's Britannia Pier, built in the 1850s, was demolished, and a new structure erected in its place. With an overall length of 810ft, the pier featured a large pavilion erected at a cost of £15,000. The structure added 'a much needed source of pleasure for the numerous visitors to this bracing resort', commented one reporter.

At Cowes, a small pier with landing stage and pavilion was built in 1901–2, and a similar structure was erected at Minehead in 1901.

Perhaps the last of the great pleasure piers, however, was the Weston-Super-Mare Grand Pier of 1903–4. It was a large and elaborate structure carrying a theatre and pavilion, plus kiosks. Financed by a group of Cardiff businessmen, it cost £120,000.

However, it was in Lancashire, where pleasure piers had their infancy, that the pier building age finally came to an end. In 1909 the Fleetwood Victoria Pier Company was formed and 30,000 £1 shares sold. The prospectus stated:

> . . . at the end of the pier will be a bandstand, windscreens, kiosk, sea water swimming baths at all states of the tide, to be constructed on the pier, every thing tending to raise our town to the position nature intended it to occupy among the watering places of Britain . . . the town and its people would be enriched by its existence.

The town and its people were enriched the following year.

EARLY TWENTIETH-CENTURY PIERS

25 Southwold (1900), was built primarily to service pleasure steamers. The photograph is c. 1905

26 Lowestoft
Claremont Pier
(1902–3), was erected
for the same purpose,
as is shown in this
1905 photograph

Changing roles

The reasons for pier building ceasing in the first decade of this century are manifold. Firstly, it might fairly be said that the coast had become saturated with pleasure piers: every resort which could require a pier had got one. Brighton had two, as had such places as Yarmouth, Morecambe, Hastings (with St Leonards), Weston-Super-Mare and Colwyn Bay (with nearby Rhos-on-Sea). Blackpool reigned supreme with three. True, piers had not penetrated into the increasingly popular county of Cornwall, but this was due mainly to constructional difficulties and the abundance of natural harbours rather than any apathy on the part of the residents. There were few other parts of England or Wales however (and certainly not the populous areas) which were not within striking distance of a pier. The people of Lancashire and the North-West had a vast store of piers within easy travelling distance. Those in the South-East were equally well served. Only the Scots seemed unable, or unwilling, to join the age of the pleasure pier. There had been proposals to remedy this. *Engineering*, dated 22 December 1899, carried the following item: 'At Ayr, on the Firth of Clyde, a company to be incorporated intends to construct a promenade pier 250 yards long, with pavilion, etc., and it will be interesting to see how this innovation from English watering places fares with the stolid Scotch character.' But like the proposed piers at such places as Brightlingsea, Filey, Seaford and Whitley Bay, the structure existed only on the drawing board.

Secondly, the financial problems associated with many piers acted as a deterrent to further investment. In the pier mania which infected the seaside at one time, optimistic claims were made which were not reflected in their profitability. Some piers never made a penny for their owners. With increasing costs of both labour and materials, pier investment was no longer the attraction it had been in the late nineteenth century. The average cost of a pier in the 1860s was about £20,000. Weston-Super-Mare Grand Pier (admittedly much larger and more sophisticated than the early structures) cost six times that amount in 1903–4.

By the early years of this century personal mobility was undergoing changes. Just as the railways had opened up the seaside to the nineteenth-century Briton, so the roads and motor transport revolutionised travel in the twentieth century. Charabanc trips were well established by the first world war, and the ease with which a holidaymaker could visit several different places in the space of a day must have detracted from the appeal of a day on the pier and the beach. The novelty and innovation which had once drawn people to the pleasure steamers at the pier head was vested in the early motor coaches.

The spectacular losses of several notable piers to storm and decay must also have militated against pier construction. The existing pier operators, in fact, had their hands full maintaining their structures to the necessary standard and in adapting both their policies and their buildings to keep pace with changing public demand.

The older established pier entertainments were being augmented by newer features. New devices were introduced to amuse and entertain the pier patrons. Slot machines, selling a wide variety of goods, appeared on pier approaches and within the buildings. 'What the butler saw' machines were another diversion where, for the usual penny, eager males would see a jerky and slightly risqué film, which experience was all part of growing up.

Motion pictures of the more conventional kind came to other piers. Eastbourne's first cinema was on the pier and after the demise of Coatham Pier the surviving landward pavilion was converted to this use.

The entertainments provided varied from one pier to another, and from one decade to another.

Folkestone Pier pavilion had passed to the control of Messrs Keith Prowse soon after the beginning of the century. They brought famous artistes to the pier, as well as Herr Moritz Wurm and his Blue Viennese Orchestra to replace an earlier Red Viennese Band. But like the other lessees before them, the operators still found difficulty in making the pier pay. There was no winter season, and takings in the summer proved disappointing.

The lease changed hands again in 1908, and two local businessmen took over the entire pier. They adopted a different attitude to pier management, seeing novel and low key attractions being more profitable than the fashionable and prestigious entertainments provided hitherto. The first international beauty show is reputed to have been held on the pier with French 'talent' being specially imported from Boulogne to compete with the local beauties. There were dog shows, strong men, dance contests, and wrestling contests. A roller skating rink was laid down near the pier entrance to attract still more visitors. Motion pictures came to Folkestone Pier, and the controversy over Sunday performances had to be settled in court. The magistrate ruled in favour of the pier operators, and Sundays in Folkestone were never the same again.

Wellington Pier, Great Yarmouth, ran the gamut of pier entertainments in the first fifty or so years of this century. Early concert and variety shows having such varied names as 'Duds Dreary Drolleries', 'Vagabond Players' and 'Royal Purple Poms' appeared on weekly bills, but their popularity prompted the engagement of concert parties for the whole season in the twenties. The first such was Dan Everard's 'Modern Follies' followed by Fred Beck's 'Modern Follies' and by Maddox's 'Evening Follies'.

Less conventional entertainments on the pier were great firework displays, speedboat trips, community singing, whist drives and roller skating. The Wellington Pier Athletic Club arranged children's races between the wars.

Visitors to the twin resorts of Hastings and St Leonards in August 1911 had a wide choice of entertainment according to the 'Public Notices' in a local newspaper of the day.

General Manager—Mr WALTER MAXWELL

TODAY (THURSDAY) AUGUST 10th.   Three Nights at 7.45 and Matinee
Saturday at 2.25.   The Great Comedian (*from Drury Lane and Daly's
Theatres*)

GEORGE GRAVES

in

"KOFFO OF BOND STREET"
Supported by full Company of London Artistes

* * *

SUNDAY AUGUST 13th.   Morning at 11.30.   Evening at 8.
MILITARY BAND. 8.45   ANIMATED PICTURES.   All seats 2d.

* * *

MONDAY AUGUST 14th.   Three nights at 7.45 and Matinee Wednesday
at 2.25.   Mr CHAS. KENYON and London Company in

"THE PRISONER OF ZENDA"

NOTE:

Prices of Admission to the Pavilion. Orchestra Stalls (Numbered and Reserved), 2s. 6d; Stalls (Numbered and Reserved), 1s. 6d; Centre Seats, 1s; Back Centre and Side Seats, 6d; Balcony (Smoking allowed), 1s. (may be booked at 1s. 6d.)

* * *

DANCING in the Shore Pavilion every Evening at 8 p.m. Admission 6d.

* * *

ANIMATED PICTURES. Morning at 11.15, Afternoon at 3.15 (When
Pavilion is not otherwise engaged).
Every Evening outside at 8.30

* * *

GARFIELD BARRATT'S CONCERT PARTY daily at 11.15, 3.15, 7 and 8.15

* * *

BATHING from Pier Landing Stage every day from 6 a.m. to 1 p.m.
High Diving; Crystal Tank Performances.
Box Ball Alleys.   Shooting Saloon.   Joy Wheel.
Photo Groups.   Afternoon Teas, etc.

* * *

Further along the seafront entertainments of a somewhat different nature were being offered.

PALACE PIER, ST. LEONARDS
Tonight at 8, Saturday at 3.   The Great London Comedy Success

"COUSIN KATE"
Seats: 2s., 1s., 6d.

Thursday at 8, in the Rink Pavilion.   Mr Phillip Goepel's Celebrated children's orchestra.   Sixty Performers.   Vocalists:
Miss Artye Crouch and Mr Jack Cornelius
(Humorous Vocalists)
Seats 1s., 6d. and 3d.

* * *

Saturday at 3 p.m. in the Rink Pavilion—exhibition of beautiful children Handsome Prizes for those adjudged the most beautiful. Children Competitors free; Visitors 2d.

* * *

Saturday, 8 till 11.   evening confetti carnival and open-air fete.
Premier Cinque Port Band

Sunday 13th August.   imperial viennese orchestra at 8.15.   Conductor Herr Wurm.   Admission 3d.   Seats 3d., 6d and 1s.

Skating daily (Thursday Evening and Saturday Afternoon excepted)
11, 2.30, 7.30.   Admission and Skates 6d.
american bowling alley open all day.

There were other entertainments. At Cleethorpes, the pavilion lacked heating and activities were normally confined to the summer months. Until 1923, the building was used principally for concerts of a very high standard; thereafter celebrity concerts were confined to Sundays, with dancing being the main attraction during the week. For some time, the pavilion was the home of the Cleethorpes Musical Festival which was highly thought of in musical circles. The provision of heating allowed dancing to take place during the winter season.

The older, well-established amusements did not lose their appeal. Pierrots still performed in some resorts right up to the second war, and there were divers still thrilling crowds at the same time.

Paddle steamers continued to give great pleasure to the new generations of holidaymakers.

On the other hand, the interwar period saw more people than ever able to enjoy a holiday by the sea, and the piers benefited from the custom. The seaside post-card often portrayed the local pier and millions of people were regaled with the classic 'Wish you were here' message on the reverse of a photograph of a pier.

There were one or two abortive pier schemes between the wars. At Hove, plans to build a pier appear to have been first mooted in 1911, and proposals for such a

PIERS BEFORE THE
GREAT WAR

27 *left* Hastings, 1911;
a wide range of
entertainments

28 *above* Blackpool
North, 1912, with the
pier packed to
capacity

structure continued into the twenties. These died a natural death in 1929, but three years later the Hove Pier Bill (apparently for another pier) was presented to Parliament. Again, nothing came of these proposals.

The alterations and improvements to existing piers, however, annually went ahead unhindered. The pier at Bognor, erected in 1865, had become the property of the local council who, in 1910, sold it to a private company. They at once spent £1,500 on it. The pier was widened at the shoreward end for nearly a third of its length. On it were erected an arcade of shops with tea and refreshment rooms attached to a theatre. At the seaward end of the pier a pavilion was erected with facilities for roller skating.

At Worthing a few miles further east, improvements were also made. Worthing Corporation bought the pier for £18,000 in 1920 and in 1926 opened a concert pavilion at the shoreward end. The total cost, including furnishings was £40,000. On 10 September 1933, fire destroyed the pavilion erected in 1887 at the southern (seaward) end of the pier. Within two years, however, a new southern pavilion had been erected to take its place, the cost this time being £18,000.

Advancement was not only confined to the south coast. Following the destruction of the 'Indian Pavilion' by fire in 1921, a new pavilion was opened in Blackpool in 1923. It incorporated seating for 1,500 persons and was used for summer revue and Sunday celebrity concerts as well as being the home of the North Pier Orchestra. In 1932 changing tastes were again reflected in the demolition of the pierhead bandstand and the erection of a sun lounge on the site. Five years later a modern bar, café and sun lounge replaced the old pierhead bar and refreshment rooms.

Something of the atmosphere of a seaside pier in the twenties is portrayed in a

29 Promenading on
Weston Grand Pier,
c. 1920

30 *opposite* Open air
dancing on Southsea
Pier, c. 1920

letter by Selwyn B. Allen published in the *Blackpool Post and Gazette* in the
mid-1970s.

My memories are of an August week in the twenties when the sun shone
brightly and it seemed the world and his wife traversed the boardwalks of the
North Pier. There I went with two holiday companions whom I had met at the
digs in Barton Avenue.

With Kathleen on one arm and Edith May on the other I spent some part of
each morning there. We felt we had proprietorial rights for Edith May's grand-
father had left her the bulk of his fortune in North Pier shares and we agreed
she owned at least a kiosk or a gilt onion dome.

Sometimes we went on the jetty or for a sail round Morecambe Bay aboard the
Bickerstaffe. At other times we would listen to George Birmingham's orchestra
in the open-air enclosure. He was quite a dandy with a mane of silver hair,
grey, swallow-tail coat and sporting a rose in his buttonhole.

We never left the pier without a visit to the Arcade Pavilion where Mr.
Jacobs and his Rumanian Orchestra played selections from musical comedies,
etc. The girls were intrigued by the way he always bowed to the audience,
especially to the handsome woman who sat apart in splendid isolation decked
out in an opulent assemblage of stone-marten furs, a tricorn hat and long jade
ear-rings. For her he removed his pince-nez glasses, bowed low and beamed
benignly.

Edith May felt there was romance in the air, especially on the Saturday
morning when Mr. Jacobs gave as an encore a piece he had just composed,
called *Because of You.*

Alas, when he acknowledged the applause the lady was missing.

I still remember the words and the tune:
> 'The sunshine seems brighter because of you,
> My sad heart is lighter; the sky more blue,
> No dark clouds can gather where hearts beat true,
> And love reigns forever because of you.'

The pavilion which saw such happenings unfortunately suffered the same fate

31 Clacton Pier in the twenties, with up to the minute entertainments

as its predecessors when it was burned down in June 1938.

Other piers were as popular. At Sandown, a new pavilion was built in 1934 to replace the makeshift arrangements consisting of a temporary stage on the sands. It immediately became an important part of the pier and of the town itself. Pier postcards of the 1920s and '30s invariably show the typical seaside resort with the roads congested with cars and coaches, the beaches crowded with bathers and deck chairs and the piers packed with holidaymakers. One or two postcards also show a different aspect of piers—their demise. The usual problems facing pier operators were just as aggravating in the twentieth century as they had been in the nineteenth. Dover Pier was demolished in 1927 after becoming increasingly dilapidated after the first world war. Ramsgate Pier suffered the same fate three years later. Towards the end of the 1914–18 war, this pier had been almost completely destroyed by fire, and shortly afterwards a drifting vessel further damaged the pier. In 1918, a mine was washed up beneath the pier and exploded. The result was a decaying gaunt skeleton which disfigured the sea front for some years. When the fifty years lease (at 5s. p.a.) expired in 1929, possession of the pier passed to the Ministry of Transport which demolished the eyesore in the following year.

Additionally, the period up to 1939 saw a large number of often spectacular pier fires. In 1904, Southsea's South Parade Pier was badly damaged by fire, although the structure was rebuilt and reopened in 1908. The year before this fire, Clee-thorpes Pier's pavilion was burned down, and the year after South Parade's reopening the first of a series of fires damaged the Great Yarmouth Britannia Pier, the others being 1914 and 1932. Other notable fires were Morecambe West End and Ramsgate (1917), Paignton (1919), Colwyn Bay (1922 and again in 1933),

32 Worthing Pier in the early thirties; crowds await the steamer

Shanklin (1927), Herne Bay (1928), Penarth (1931) and Morecambe Central (1933). One fire which made the headlines throughout the country was that on Hunstanton Pier on Sunday, 11 June 1939. Two women were sitting at the end of the pier when they noticed smoke issuing from the pier buildings. Within minutes, the pavilion was in flames and the two holidaymakers were forced to jump thirty feet into the sea below, where they were taken to their boarding houses suffering from shock. Regular and auxiliary firemen from Hunstanton and Old Hunstanton prevented the fire from involving the rest of the pier, but the concert hall, café and waiting rooms, together with the entire equipment of a touring concert party, were lost. Dramatic newspaper photographs of the two women leaping from the burning pier appeared on breakfast tables throughout the country the following morning; the number of people who consequently decided not to patronise the pier that year is not recorded. Within a few months of this fire, however, a more serious threat presented itself.

Piers at war

The commencement of hostilities in September 1939 brought a change to the seaside. In many areas the sea front was fenced off, and beaches were festooned with barbed wire and defensive paraphernalia such as stakes, girders and pill boxes. Piers were singled out for special treatment.

It was considered that piers could form excellent landing stages for invading troops, and it was decided to arm some, and to breach these in threatened areas, that is the south and east. Military demolition teams arrived in many resorts in the spring of 1940. At Eastbourne such a squad started placing explosive charges during the performance in the pier theatre. The staff were given three days to remove as many items as possible from the pier, but eventually the planned demolition of part of the pier was not carried out, and part of the decking was removed. Cromer pier was similarly treated, but as this action removed the only means of reaching the lifeboat station at the pier head, the gap had to be bridged with planks.

Other piers were less fortunate. Bournemouth, Cleethorpes, Deal, Folkestone, Redcar and Ventnor piers were but a few of those where sections were blown out.

33 Worthing firemen
damp down the
remains of the
pavilion and pier head
in September 1933

The piers on the west coast were on the whole spared this fate, although Mine-head Pier was completely demolished to provide a clear line of sight for nearby gun batteries.

The hazards facing piers were magnified during the war. Enemy action, drifting mines and stricken ships were added to the usual dangers. Deal Pier was particularly unfortunate, being both breached by the military and by a mined vessel in 1940. Plymouth and Eastbourne Piers were also both damaged by enemy action.

The breaching of piers had other consequences. When the pavilion and pier head at Folkestone caught fire in May 1945, firemen had difficulty in reaching the fire and fighting it.

Southend Pier occupied a unique place in wartime pier history. The Royal Navy took the pier over in August 1939 as the Thames and Medway Control headquarters and renamed it 'H.M.S. Leigh.' It was heavily fortified with pill boxes, anti-aircraft guns and depth charges. The only serious attack on it, in November 1939, by aircraft of the Luftwaffe, was successfully beaten off. The pier's main role was as a convoy assembly point and the first such sailed from there on the 7 September

34 Deal Pier after a damaged vessel sliced through it in 1940

1939; a further 3,300 convoys were to follow. Not only was the pier head used as a conference and operations centre, but the structure was used to load men and supplies on to the ships. The pier railway ran day and night to carry the sick and wounded from the ships and to replace them with fresh men. Food and amunition were carried a mile out to sea in the electric trucks. The pier's main supplied fresh water to vessels mustering nearby. Southend Pier was demobilised, along with many of the servicemen it had so faithfully served, in 1945.

The postwar era

The end of the war and the release of piers by the military left many operators with sad and dilapidated structures which bore little resemblance to the gay, crowded piers of prewar days. They were left in something of a dilemma—reinstate and re-open or abandon and demolish. There were, of course, many factors to be considered. The state of repair of the structure and the likely cost of repair were the major points, although war damage compensation was available from the State.

The attitude and determination of the owners, the likely returns from the pier and the cost of demolition also had a bearing on the matter.

In many cases, repairs were effected and the structure re-opened with minimum delay. Southend was opened by May 1945, and Bournemouth by August the following year. Worthing Pier received its first postwar visitors in April 1949.

Elsewhere it was decided that restitution was not feasible, and demolition the only practical course. Such illustrious structures as St Leonards, Folkestone and Plymouth Piers were lost in the immediate postwar years.

The fire on Folkestone Pier had left the structure a blackened eyesore but repeated efforts to remove the structure met with legal complications. Finally, in 1952, demolition started and there were only a few people who mourned the passing of the pier. At St Leonards whatever hopes there might have been for reinstating the pier were dashed by severe gale damage in March 1951; the remains of the pier were demolished the same year.

A few piers, although doomed, lasted a few years more. Lee-on-Solent Pier survived until 1958 and that at Lytham until 1960.

Elsewhere a compromise was reached. Part of the structure (often the truncated section seaward of the breach) was demolished and the remainder utilised. Such piers as Cleethorpes and Redcar were left with little more than the pier pavilion and a short promenade, but even in this shortened form they continued to play the part of the pleasure emporia.

In one notable case, the pier was demolished and a completely new one erected in its place. This was at Deal, where the pier of 1863 was found to be beyond repair; the local authority nonetheless considered a pier to be essential to the successful development of the resort and consequently a new pier was built in 1954–7 at a cost of £250,000.

Other major reconstruction work included that at Ventnor. Here some of the original structure was incorporated into the rebuilt pier, which featured a pier-head entertainments centre with sundeck, bar and bandstand. The 'new Royal Victoria Pier' was officially opened on 28 May 1955.

Those piers which escaped the demolition gangs' torches manifested the varying fortunes in the 1950s and '60s which had characterised piers from the early days. A number of problems were common to all piers: in the sixties, domestic holidays were facing severe competition from foreign tours, particularly cheap air holidays to the Mediterranean area. Many of the habitual seaside holidaymakers were being syphoned off by these new experiences in travel. Many of the newer generation had not been brought up in the tradition of a week by the sea and owed no loyalty to the resorts. Compared to the prospect of a fortnight in the sunshine of a foreign country, surrounded by new experiences, sights and sounds, a two-week holiday in a deck chair at the end of a pier, listening to the orchestra playing familiar tunes, must have seemed rather passé. The piers suffered with their resorts.

Some piers made determined efforts to win back their customers. The old-fashioned kiosks, bars and cafes, with dull paintwork and wooden floors were swept away to be replaced by purpose-built erections of modern design with comfortable furnishings and wall to wall carpeting. The three piers at Blackpool were extensively refurbished and modernised in the fifties and sixties. Smaller piers were also the subjects of such developments.

At Mumbles, the pier was re-opened in 1956, complete with new landing stage. At the official ceremony in June of that year, a new regular pleasure steamer service from the pier was inaugurated. Ten years later, the structure's appeal was

35 The rebuilt and
refurbished Ventnor
Pier, opened in May
1955

further enhanced by the provision of an amusement arcade at the shore end.

Other authorities saw fit to increase the facilities on their own piers. The local authority in Bournemouth provided a new pier-head theatre on Bournemouth Pier in 1960, and a roller skating rink with amusement arcade on Boscombe Pier two years later.

By the sixties, Clacton and Walton Piers had acquired the amusement hardware for which they are noted. Just about every conceivable facility was provided and the immense popularity of the structures fully justified the changes made.

Elsewhere, pier operators were not so fortunate. The severe storms of 1953 damaged a number of east coast piers, particularly Cromer. In a storm on 3 March 1965, the pavilion at the seaward end of Bognor Pier collapsed into the sea, and, except for the entertainment complex at the promenade end, the rest of the pier fell into disuse.

There were lighter moments. Hunstanton Pier starred in a comedy film of the fifties entitled *Barnacle Bill*, a dilapidated Victorian pier in the fictitious resort of Sandcastle. The new owner, faced with hostility from the local councillors who all had a vested interest in the demolition of the pier, breached it at the landward end and registered it as a cruise ship, the *Arabella*, with the 'obscure country' of Liberamia. The film carried the idea of a pier as a luxury liner, which never left the beach, to its ultimate conclusion and the idea may, momentarily, have provided food for thought for genuine pier owners.

Unfortunately, the real Hunstanton Pier was not as successful as the structure it portrayed in the film. After the war the pavilion was not rebuilt, and for a time the pier enjoyed popularity as a roller-skating centre, and also boasted a zoo and miniature railway. Most of the pier, however, fell into disuse and only the shoreward end remained in regular service. In 1964, a two-storey amusement centre, ballroom and restaurant replaced the old amusement arcade and café.

But problems beset both large and small piers. The Southend local authority, owners of the world's longest pier, were faced with a proportionately large problem. In 1949–50, the pier had a record of 5,750,000 visitors; by 1969 admissions had fallen to only 1 million. The fleet of thirteen vessels which served the London–Southend route had disappeared by the late sixties, and in 1969–70 the pier made an overall loss of some £45,000.

The end of the 1960s saw such piers as Southend endangered by increasing costs and decreasing patronage. This period also saw some piers closed to the public whilst owners, local authorities and local residents debated their futures. It also saw a number of piers admirably fulfilling their roles as popular, useful and attractive profit-making enterprises. The 1970s, however, were to see more public interest in piers than at any time since the pier age of the previous century, this when piers were facing more problems than ever before.

36 The abutment of Rhos-on-Sea Pier, demolished in 1954. The pier had been originally built at Douglas in 1869, and re-erected at Rhos in 1896

MEN, MATERIALS AND METHODS

Introduction

Britain led the world in nineteenth-century civil engineering, and seaside piers
constituted a small, but important, part of the industry's output. Techniques and
materials tried and tested in pier construction found important applications in
the many and varied structures erected by British engineers and contractors
throughout the world.

It is as well to note the basic idea behind pier design. Although solid masonry

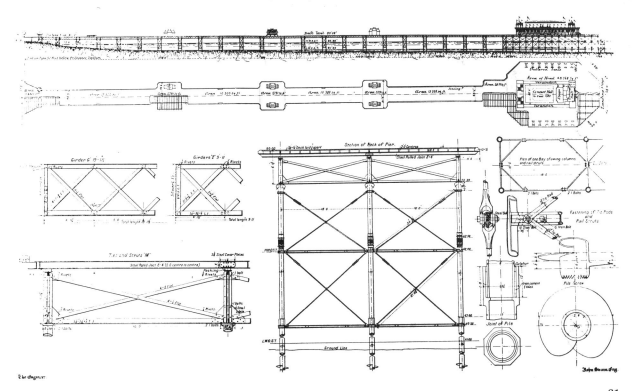

THE NEW PIER AND MARINE PALACE, BRIGHTON.—GENERAL VIEWS AND DETAILS

MR. R. ST. GEORGE MOORE, ASSOC. M. INST. C.E. ENGINEER

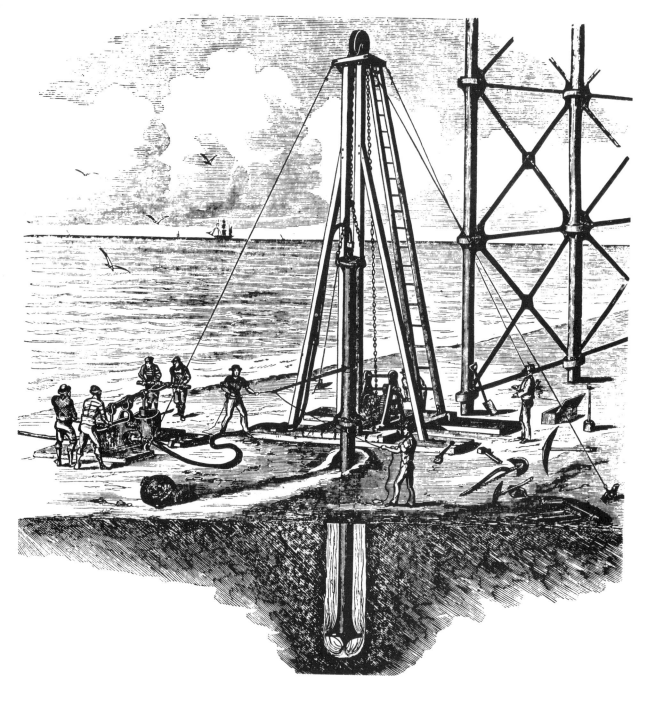

38 'Jetting' of piles during the extension of
Southport Pier, 1868

piers had been in use since time immemorial, piled construction was adopted for seaside pier construction for four main reasons:

1. it was considerably cheaper than masonry construction;
2. it offered little resistance to wave motion and so made vessel mooring easier;
3. it did not much alter movement of solid materials (sand, pebbles, etc.) which might otherwise change the character of a whole section of beach;
4. it became an inherent part of a pier's appeal.

Generally speaking, a piled seaside pier consisted of four main parts, which may be described as follows:

1. the piles or columns, which were the uprights carrying the structure;
2. the girders or horizontal members spanning the piles at the top;
3. the decking, surmounting these components;
4. the various struts and braces providing rigidity throughout the other three parts.

These factors were all bound together in the overall design which envisaged a pier 'head', usually rectangular, built in the sea, and connected to the shore by the 'neck'. Intermediate widenings of the neck might also have been provided. It was on the head and other widenings that the pavilions and other pier buildings were usually erected. The only two exceptions to this rule were the suspension piers at Brighton and Sea View where the necks were in effect suspension bridges; for various reasons this design was not widely adopted. In nearly all cases, landing stages were provided alongside piers to facilitate passenger landings from steamers at different states of the tide.

Early efforts

In the early days of pier construction, timber was almost exclusively used. Timber at that time was readily available and its manipulation understood and easily accomplished by the engineers and contractors of the day. Oak, fir, beech and pine were popular choices, and the structures were feats of joinery. The piles, driven into the sea bed by piling engines, were joined by stout timber beams and deck pieces, and strengthened by walings, all securely jointed or spiked together. Some piers, such as that at Ryde, were of relatively simple design although ex-

40 Details of the pier structure, prepared by
Clarke & Pickwell, 1880

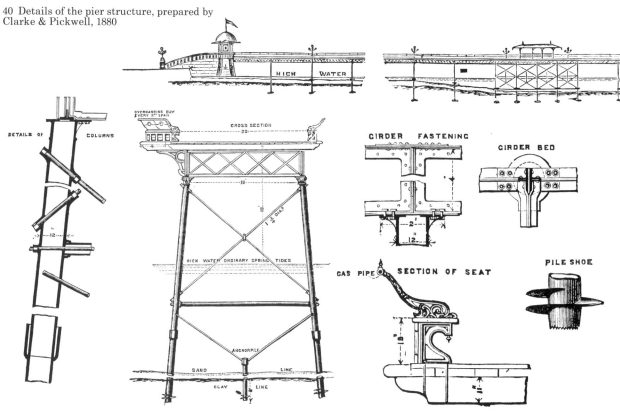

tremely long; the pier's original length of 1,250ft was continually increased and a
pier head later added, to give a finished length of over half a mile. Telford's Herne
Bay Pier of 1831–2 had an overall length of 3,600ft with a T-head, and featured a
crane-operated incline at the seaward end. Grandiose schemes for a stone pier
head complete with a colossal statue of William IV in naval uniform, concealing a
refreshment room and lookout point, never materialised.

Brown's Brighton Chain Pier of 1822–3 employed a different principle. Basically
there was a pier head and three intermediate platforms between it and the shore.
All were carried on wooden piles and supported massive cast iron towers. Between
these were slung suspension chains from which the pier deck was suspended. At
the shoreward end the chains were buried in tunnels excavated deep in the cliff
face, whilst at the opposite end the chains were securely anchored to the pier head.
Unfortunately, the pier was neither a financial nor structural success. The suspended deck was highly susceptible to wind damage and repeated modifications
were required to repair and strengthen the structure. Only one further suspension
pier was erected, and various design features incorporating lessons learned at
Brighton ensured that the Sea View Pier (Isle of Wight) was a more successful
structure.

At first, these early piers were hailed as great engineering successes. They had
not been easy to build. Difficulties in driving the piles on the Chain Pier had
caused the contractor to give up in despair, and the work was completed by the
designing engineer using direct labour. The problems did not end with the
completion of the piers. Within a few years serious defects became apparent in a
number of structures. Domestic timbers used in a marine environment are prone
to attack by the marine worm and borer—the *limnoria terebrans* and *teredo*

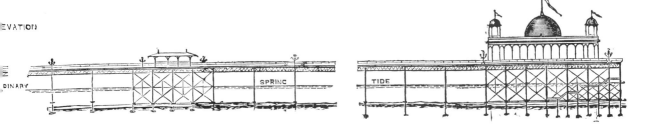

41 Details of the superstructure and buildings, 1880

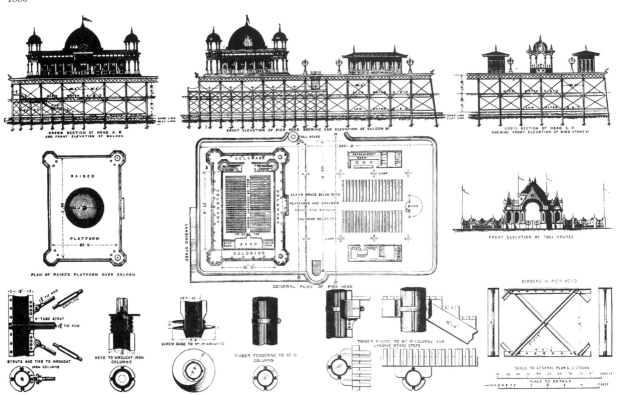

navalis. These animals eat and bore their way through timber in a very short time, and by 1839, seven years after its completion, Herne Bay Pier succumbed. It was so badly damaged by 'worm' attack that it was pronounced unsafe and its closure recommended. A close inspection of the structure revealed some alarming facts. Some of the piles were completely severed, and the upper parts hung from the deck instead of supporting it. Other piles were reduced in thickness from the original 12–15 inches to as little as 3 inches and were virtually useless. Forty-two 'crutch' piles were placed to support the weakest parts of the pier, and two years later extensive repairs were carried out. The outer piles were replaced with new wooden members protected by scupper nails whilst on the inner parts cast iron piles were used as replacements.

At Southend much the same thing happened. There, the pier head was built first, and the neck was to be constructed later as funds permitted; in the meanwhile, the head was used to tranship passengers from steamers to carts and rowing boats.

The piles in the pier head had been protected with nails (to form a protective layer against the 'worm') but in 1842, four years after its completion, the structure was literally on its last legs. Drastic action was clearly necessary and in 1844–6 the pier extension was built, consisting of a new head and neck connecting this to the old pier with a total length of one mile. For economy reasons, the first third of the structure was of protected timber; the piles were exposed at low tide and it was supposed that frequent inspections and repairs would present no difficulty. For the rest of the structure, however, cast iron was adopted. There were $13\frac{1}{2}$in square piles acting as 'shoes' to the fir piles, and the hollow parts were filled with concrete and sand to prevent the ingress of water.

The success of these iron piles, where they were used in conjunction with wood, prompted the development of all iron piles. Cast iron is extremely strong in compression and, if of high quality, highly resistant to corrosion. It was thus eminently suitable for pier piling.

Iron takes over

The first development was the casting of circular piles which were easier to place and, once in position, offered less resistance to wave action than the square members. These cylindrical castings were fitted with flat base plates to carry the weight of the superstructure and superimposed load; these were placed by excavating holes in the foreshore, lowering the piles in to them, and then back-filling.

These piles were an improvement, but the disturbance to the ground led to some instability; an engineer came up with the answer. Mitchell's screw piles consisted of screw blades fitted to iron castings which were literally screwed into the ground. Earth pressure acting on both the top and bottom surfaces of the blades as well as frictionally on the pile body ensured a rigid foundation.

At first, it proved difficult to connect the continuous rigid lengths of iron in straight lines and to keep them in plumb. Consequently, the piles were thereafter cast in two sections—the lower with the screw attached (often known as the piles) and the upper (usually termed the column). Sometimes, the upper part of the pile was enlarged to form a socket into which the lower end of the column sat. If necessary, further components were joined vertically by a similar socket joint. In other cases, each component terminated in a external lip; these were locked together to form a flanged joint.

The first application of screw piles in this country was made at Margate Jetty in 1853–6. The piles were placed by holding them upright in gantries and rotating them by means of capstans. In the early days, men or horses turned the capstans but later steam engines were employed in this capacity.

Screwed piles were not exclusively employed, however. Driven piles were favoured by a number of engineers but the heavy weight of the 'monkey' being dropped on to the pile in the piling engine often fractured them. Again an engineer came up with a solution. Dixon introduced the dolly, a wooden and india rubber block which sat atop the pile during driving and absorbed the shock. But Dixon's ingenuity did not stop here. In certain ground conditions, neither screwing nor conventional driving provided a satisfactory foundation, and for the sandy shore of the Lancashire coast, this engineer perfected the jetting technique, originally introduced by Brunlees in the 1850s.

The piles used in this application were thinner than those used in screwing since the piles did not rotate and did not therefore have to possess high torsional

LATER PIER
CONSTRUCTION

42 Contractors Head
Wrightson at work on
Herne Bay's third pier
1896–9. The lines for
the steam crane were
later used for the pier
railway

43 Hydraulic
screwing apparatus
used in pier con-
struction towards
the end of the
nineteenth century, as
depicted in a
contractor's catalogue
of 1890. top: Plan;
bottom: Elevation

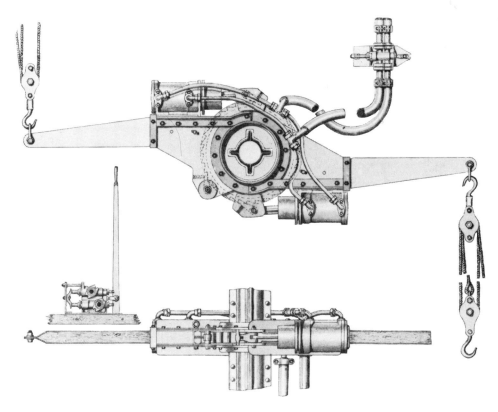

(twisting) strength. They were, however, open at the bottom end and fitted with a serrated base plate. In positioning, the pile was held in moving guides in an ordinary piling engine and a tube passed down its centre. This was attached to a small hand-operated pump or domestic water mains, and water forced down the pile and into the sand beneath. The agitated sand boiled up and the pile sank into it, being assisted by labourers twisting it slightly back and forth. As soon as the pile was placed to a sufficient depth, the pump was knocked off and natural compaction of the sand held the pile firmly in place. This method was used in the Southport Pier constructions of 1860–8. Piles were all sunk 15–20ft in an average time of 20–30 minutes, some being sunk from a raft where the beach was not exposed at low tide. A slightly faster rate of progress was achieved on Blackpool Victoria Pier in the nineties where the contractor Finnegan used a steam fire pump to provide the water pressure.

Other forms of piles were used (including segmental cylinders and pseudo-elliptical piles formed of Barlow rails bolted back to back) and other methods used to place them. In rocky foreshores, the old established method of excavating piles and backfilling them continued to be employed, although the piles were often bolted to the rock and secured in place with concrete.

Struts and girders

Concurrent with the development of the piles was the progress made in the design of the girders spanning the piles and the struts and braces which ensured rigidity throughout the structure.

At first, timber continued to be employed for the joists spanning the piles and a number of designers continued this feature well into the pier age. However, iron soon began to eclipse it. Cast iron was not a success in this role since it does not possess high compressive or torsional strength. Wrought iron, on the other hand, found an ideal application here. The actual design of the girders varied from pier to pier, and from designer to designer. In some cases, solid plate girders of I-section were employed and in some cases these fulfilled the secondary function of forming seating along the pier. Elsewhere, various types of lattice girders were used. They had the advantage of containing less iron, and therefore usually being lighter. They also gave greater scope for design. Various types, rejoicing in such names as N-trusses, Pratt girders and Warren girders, were employed, but they all basically consisted of a combination of horizontal/vertical/diagonal members taking the various forces acting on the girders.

The bracing to the piers was of the utmost importance. Continual battering by waves and high wind forces presented the very real possibility of collapse like a pack of cards if additional struts (in compression) and ties (in tension) did not strengthen the structure. Horizontal bracing was often angle irons or rails, whilst diagonal members were often wrought iron rods, adjusted by turnbuckles to give the correct tension. They were attached to the piles by iron collars.

Whilst the rest of the piers changed, decking was almost always of wood, although both timber and iron were interposed between the deck and the girders atop the piles. Decking was often laid in a herring bone fashion to impart extra strength to the pier, and it was usually cambered to allow easy run off of water. In some cases, gaps were left between the decking pieces for the same purpose, and to allow high waves to pass through the deck to prevent damage.

Timber was almost exclusively used in the landing stages surrounding the pier

44 Pier construction
in the 1950s.
Contractors Concrete
Piling at work on
Deal's third pier in
1956. Reinforced
concrete was used
throughout

PIER SUBSTRUCTURES

45 The substructure
of Birnbeck Pier,
Weston, with
interesting bracing
to the columns. Birch
was the engineer

46 Typical pier substructure – raked and vertical piles support lattice girders on Eastbourne Pier. Later additions have masked much of Birch's work

heads. Shocks from vessels striking the structures when mooring required both strength and resilience, and the frequent necessity for repairs also required cheapness. In some cases, however, ironwork was employed, but with substantial protective fendering.

Methods

The physical means of constructing piers required both skill and ingenuity. Basically pier construction followed a sequence. Falsework was built out from the promenade as a working platform or else moored barges provided a suitable base. The piles were placed and once a group was in position, girderwork was added, bracing attached and the decking laid, and the scaffolding and machinery moved further out to sea. Things were seldom so simple in practice.

The builders of Clevedon Pier employed piles consisting of pairs of Barlow railway lines riveted back to back, supported on screw piles and on columns concreted into holes in the rock. The difficulties they faced were daunting. A tidal range of 45ft and currents of up to five knots concealed hard magnesium limestone as well as soft slushy mud. A pontoon and barge used in the work were smashed by the seas, but a third barge survived the course. The piles and columns were placed by temporary derricks on falsework construction in the water. The girders were floated out on pontoons and winched atop the columns. Progress varied with the weather. Sixty men took ten days to erect one span, and three months to build another.

Rather more difficulties existed at Westward Ho! Pier. Poor sea conditions ruled out the use of pontoons and staging had to be built out from the sea front and, as the work progressed, from the completed part of the pier. A jumping rig was used to excavate holes in the rock on the pier site. This consisted of a jumper— a winged projectile 4ft 6in long—which was dropped down girders onto a cylinder in the rock; a constant rise and fall of the jumper excavated the required hole. Again, the time factor varied. Some holes were made in twenty-four hours, whilst those constantly covered with from 6–9ft of water with a constant surf and heavy ground swell took a week or more. The piles themselves were lowered into the holes and wedged and concreted into position by a diver working through trap

doors in the sides of the iron guide cylinders.

Even in calmer positions, mishaps occurred. Storms and high seas damaged many piers whilst they were under construction—Coatham and Blackpool North for example. Months of work were sometimes lost in a single night. But there were problems in normal piling operations, perhaps the most aggravating of which was the fracture of piles despite all precautions being taken. At Withernsea Pier, this problem had three possible solutions. If the damaged pile was the first of a pair, it was abandoned and the 'replacement' pair placed about three feet away; the resultant difference in length between adjacent bays was not discernible. If the pile was the second of a pair and the fracture was near the socket, a short piece was secured to it to correct the height. In the case of piles broken off below ground level, a larger conical pile was cast and telescoped over the offending component. In all cases, removal of a damaged pile was found to be so tedious and time consuming that such a course was not adopted.

Oddments

48 Weston Grand Pier: immortality for contractors Mayoh & Haley on the cast iron columns

As in everything else, there were exceptions to the general rules. It would be satisfying to indicate a logical progression in pier design but this is not possible. Financial considerations, local sea and ground conditions, availability of materials and the designing engineers' preferences and prejudices all had their effects on pier design. At Brighton West, for example, an attempt was made to use solid wrought iron rods some 5in in diameter instead of the conventional cast iron cylinders. Within a few years, however, they were found to be so badly corroded that they had to be replaced by cast iron. At a number of other sites, on the other hand, wrought iron piles (usually in the form of Barlow rails) were used with unqualified success.

At Eastbourne, the shoreward end of the pier was destroyed by a storm in 1877 and was subsequently rebuilt at a higher level. Hastings Pier, however, incorporated a forked entrance section where a large open area was left to allow heavy seas to pass upwards through the pier to reduce the risk of damage.

There also appears to have been a certain amount of under- and over-designing. In other words, some structures were too weak for the job they had to perform and

required strengthening, or else others were so strong that they remained standing long after corrosion and accidental damage should have taken their final toll. Piers, perhaps, supported the theory that despite calculations and model testing, the only way to really test a structure was to build it and expose it to service conditions.

Further developments

The availability of good quality cheap steel in the late nineteenth century offered further possibilities to pier designers. Of more uniform quality than wrought iron, and also cheaper, steel began to supplant this material in the girderwork and joists below the decking. Like wrought iron, it was susceptible to rusting, but was usually finished with some protective coating. Cast iron continued to be used for piling. Bangor, Cromer and St Leonards Piers were just three in which steel was extensively employed.

Another interesting development around the turn of the century was the return to timber in a number of piers, including Southwold, Felixstowe and Swanage. Foreign timbers, notably green heart and jarrah, had been utilised abroad for marine works and ship construction because of their high resistance to so-called worm attack. This was generally successful in British pier work, although local sea conditions in Swanage Bay resulted in faster deterioration than had been expected.

One pier which appeared to have defied the conventions of pier construction was Great Yarmouth's Britannia Pier, built in 1900–2. In the shallow waters beneath the pier neck, steel screw piles (104 in all) were employed whilst on the head some 200 karri wood piles were driven. Further down the coast another apparent contradiction to accepted rules was the rebuilt Clacton Pier of 1890–3. Here, the extended and improved pier was mainly of pitch pine, with some steel piles beneath the head to support the weight of the pavilion and to stiffen the pier against vibrations caused by berthing steamers.

As a further precaution against the transmission of vibration, a one inch gap covered by wrought iron plates was left between the head and the neck.

Engineers were continually striving to produce more sophisticated and ornate piers, whilst at the same time cutting down on unnecessary ironwork and reducing costs. By the 1870s, pier building was such big business that it was reputedly possible to order all necessary components from an iron founder's catalogue. This may account for the similarity of components on different piers, although consistency of design on the part of the engineers must also have been responsible. Engineers were not only battling seas and storms, however. Sometimes a misplaced sense of economy and cheeseparing on the part of the pier owners resulted in inadequately built piers which occasionally failed. Piles screwed to an insufficient depth were an example, and engineering ignorance or incompetence were not necessarily the responsible factors, as has sometimes been suggested. A large head and frequent embayments on the neck appeared to ensure adequate rigidity but this did not always fit in with the pier owners' schemes.

Engineers and contractors

The men who were charged with designing and erecting these piers were a mixed crowd. Pier building produced its specialists, the most famous being Eugenius Birch (1818–84). This doyen of pier engineers was an accomplished artist and mechanic, and his early career involved railway and bridge works both in this country and in India. In 1853, he designed the pier at Margate using screw piles, as has been mentioned. Following the default of the contractor, Birch used direct labour to finish the pier. Shortly afterwards he designed one of the great wave of pleasure piers—Blackpool North in 1862-3. Birch was an imaginative and competent engineer, and the more piers he designed the more he was in demand from pier promoters throughout the country. He designed Aberystwyth, Brighton West, Deal, Eastbourne, Hastings, Lytham and Plymouth, amongst others. In engineering circles, Birch's name was synonymous with piers, as was Brunel's with bridges or Locke's with railways. He was the most prolific of pier engineers and his last structure—Plymouth—was opened in 1884, the year of his death. The West Surrey waterworks, the Devon and Somerset Railway, Exmouth Docks, Ilfracombe Harbour and the Scarborough and Brighton Aquaria were all products of Birch's fertile brain.

No other engineer could match Birch's output. Joseph William Wilson—one-time assistant engineer on the Crystal Palace construction—designed several piers, including those at Bognor and Hunstanton. James Brunlees, a noted and respected Victorian engineer, designed the first true pleasure pier at Southport in 1859-60. His sporadic output thereafter included Llandudno and Rhyl Piers. Another famous engineer who turned his hand to pier design was John James Webster who, working at the turn of the century, was responsible for such structures as Bangor and Dover.

There were a host of other engineers engaged on pier work. Many were successful engineers in their own fields who prepared plans for the odd pier as the occasion demanded—Thomas Telford, the Rennies and Robert Rawlinson, for example. Others were less well known, but had a local connection with their structures. The Bournemouth engineer Archibald Smith designed Boscombe and Southbourne Piers, whilst R. St. George Moore was responsible for the highly sophisticated St Leonards and Brighton Palace Piers.

Much the same comments could be applied to the contractors. Three names stand out. Head Wrightson of Stockton, Alfred Thorne of London and Robert Laidlaw of Glasgow took the lion's share of pier contracts. Laidlaws were in the pier business from the start, and they built several of Birch's piers, as well as one for Brunlees. After 1870, however, they seem to have been eclipsed by Head Wrightson as pier builders extraordinary; this firm entered the scene to complete piers started by J. E. Dowson. Dowson had, like Laidlaws, been involved in the early days, being contractor for Birch on four occasions, and for Wilson on Bognor Pier. Dowson died in the late 1860s whilst Eastbourne and Scarborough Piers were still under construction; Head Wrightson completed them and were soon submitting tenders to build other piers from scratch. They were awarded contracts for Redcar, Cleethorpes, Southsea and Ramsgate in the seventies; for Ramsey, Skegness and St Leonards in the eighties, and for the extensive Herne Bay Pier of 1896-9. By this time Alfred Thorne had emerged as a pier contractor, and several of the later piers such as Bangor, Cowes, Cromer and Dover were his work.

There were numerous other contractors engaged in pier work, but for the most part they built but one or two. Several were local firms who built only within their own areas—Howell of Poole for example. Many more were general engineering or railway contractors with no special interest in pier work. MacIntosh, the original contractor for the Chain Pier, was a one-time canal and public works contractor, and others (e.g. the Widnes Foundry Co.) were iron founders assembling their own products. A few ensured immortality by casting their names on many of the pier components; most were forgotten as soon as the last bolt was placed and the pier opened.

Several enterprising engineers performed the dual function of engineer and contractor, sometimes on the same pier. The brothers James and Arthur Mayoh of Manchester were contractors for Brighton Palace, Mumbles and Penarth Piers. and designed the second Great Yarmouth Britannia and Bexhill Piers. Dowson, contractor for Aberystwyth, Bognor, New Brighton, Eastbourne and Scarborough Piers, also designed Redcar Pier, built by Head Wrightson. But perhaps the most notable engineer/contractor was John Dixon. He was a noted railway builder both in this country and overseas, especially in China. He was responsible for bringing Cleopatra's Needle to England, and as already mentioned made important advances in pier construction—the introduction of the piling dolly and the use of jetting. His contracts for the Southport and Llandudno Piers were competently and speedily executed. In the case of Douglas Pier, Dixon designed the structure, supplied the components and erected it within five months! Dismantling of this pier in the mid-1890s and its re-erection at Rhos-on-Sea seem to have presented no problem.

The new century

As long as piers were being built, the repair of existing structures, as well as their extension and improvement, presented little problem. Components were readily available, and there were engineers and contractors experienced in the work. After the first few years of this century, however, these factors no longer applied. Furthermore, the introduction of reinforced concrete and the universal use of steel had their effects on pier work.

When New Brighton Pier was substantially rebuilt in 1928–30, steel was used in the form of new girders and bracing and as reinforcing collars to damaged piles. Some reinforced concrete was also used, but this was the major material in the reconstruction of Boscombe Pier's head in 1926. Reinforced concrete—concrete reinforced with steel bars—is extremely strong in tension and compression and can be made very durable in a marine environment. At Boscombe, it was used in both horizontal and vertical members, and the pier neck was rebuilt in a similar manner shortly after the war.

Reinforced concrete was also used to repair the piles of other piers. At Cromer, 57 of the 91 piles were found to be worn and even holed in 1930. Contractors May, Gurney & Co. of Norwich encased the damaged piles in concrete cylinders carried down to the chalk into which the piles were driven.

At Swanage the wooden piles were 'bandaged' with reinforced concrete jackets where the piles had been damaged by worm attack.

Reinforced concrete found an interesting application in the proposed (but abortive) Hove Pier between the wars. The plans envisaged basically a piled structure on piles at 105ft centres, supporting reinforced concrete longitudinal beams and decking of the same material. A pier theatre, kursaal and railway line below deck level in a tube formed by the inner two of the four longitudinal beams were features of the design.

Postwar aspects

The problems presented to the engineers and contractors charged with dealing with piers in the immediate postwar years were basically of three types involving the demolition of some piers, the repairs of others and large scale rebuilding of a few. Demolition was complex in that the pier had to be completely removed so that there was no risk to vessels or bathers from submerged ironwork. At Folkestone, explosives had to be used to bring down some of the columns.

At Worthing, the breach in the pier (some two and a half bays) was closed by the insertion of steel sections. However, these were fabricated to match the existing components and the appearance of the pier was not altered.

At other piers, where breaching was not carried out, re-decking often had to be done in concrete because of the shortage of timber—Eastbourne Pier was repaired in this way.

On the Isle of Wight, the repair of Ventnor Pier was not considered feasible, and substantial rebuilding took place. Using part of the original substructure, the pier neck was built of welded steel on cast iron piles, decked in green heart. The pier head was entirely of reinforced concrete.

At Deal, where an entirely new pier was decided upon, plans were prepared for a reinforced concrete structure by Sir William Halcrow & Partners; contractors Concrete Piling Ltd erected the structure over a three year period. Hollow steel piles, surrounded by a concrete pipe, and filled with concrete and grout respectively, supported a structure of steel beams encased in concrete with additional reinforcement. The decking was also of concrete. Modern materials and methods were used to produce a pier of traditional layout.

Throughout the 1950s and '60s, repairs have been carried out to various piers. At Cromer, an inspection of the pier by consulting engineers Lewis & Duvivier in the mid-1950s revealed that many of the concrete encasements placed around the piles in 1930 had been undermined and beach movement coupled with abrasion by moving sand and shingle meant they no longer rested on the chalk. Underpinning was carried out in 1955 and in 1968.

The girderwork had also suffered from the effects of spray, salt air and wind-driven sand and shingle, and the bracing had been exposed to the highly corrosive combination of submersion and exposure. A comprehensive programme of cleaning and painting was accordingly undertaken.

Similar problems occurred on other piers. In a number of cases, renewal of members with steel or concrete was required because corrosion was so far advanced that repair was impracticable. Colwyn Bay Pier was an example. In others, reinforcement was required to support new pier buildings, as at Bournemouth, where a completely separate concert substructure was inserted beneath the pier theatre of 1960.

Numerous piers had repairs effected to the piles by encasing them in concrete as at Cromer. Occasionally cylinders filled with concrete, as opposed to unprotected concrete, was employed.

Many piers consequently exhibited structural work of different styles and periods and in differing states of repair. Some members may have been provided to carry loads (such as buildings) which no longer existed. Repair and maintenance work presented difficulties of both identification and execution.

In the 1860s civil engineering was as much an art as a science. It required skill and experience on the part of the engineer and tenacity and ingenuity on the part of the contractors. Pier repair work a century later may have had the benefit of modern machinery, up to date materials and advanced techniques, but it was probably true to say that the qualities required of engineer and contractor remained substantially unchanged.

FIVE

THE ARCHITECTURE OF AMUSEMENT

It is sometimes difficult to draw the line between pier engineering and architecture. The engineering aspects—the load carrying—have been discussed. Practical and utilitarian though the substructures were, they were often decorated and embellished to an impressive degree. It was above deck, however, where flamboyance and ornateness reigned supreme. In many cases the same hand which designed the piles and girders prepared the blueprints for the delicate ironwork, painted woodwork and solid stonework of the pier buildings, kiosks and seating. Only in the closing years of the nineteenth century did the design of the piers and their pavilions become separate undertakings. The diversity of engineering works, however, was matched by the variety of superstructural components, and the latter may fittingly be called the architecture of amusement.

In all about 90 pleasure piers were built, of which approximately half survive today. Twenty-nine of these have now been listed by the Department of the Environment as buildings of special architectural or historic interest. The most famous English pier to be erected and probably the most beautiful was an early example—the Chain Pier at Brighton.

The pier was of suspension bridge construction and its chains were driven 54ft into the cliff. The building operations took less than a year, and the pier was opened on the 15 October 1823. The designer was Captain, later Sir Samuel Brown, RN, who had already built the Union suspension bridge over the Tweed at Berwick. The pier comprised four clumps of piles surmounted by four pairs of towers, of which each pair formed a kind of archway with a pediment over. These were connected to each other by a continuous suspension bridge of four sections which led to a T-shaped platform at the sea end containing baths and later a camera obscura.

It was the towers which gave the pier its special distinction. Like Brunel's later suspension bridge across the Avon at Clifton, they derived something of their character and feeling from the great gateways and temples of ancient Egypt. The Chain Pier was thus the very first example of that oriental influence which was to have such a marked effect on the style of most piers to be built in England at later dates.

Each tower contained a small shop or store. At the shore end of the pier were three cottages with delicate Regency trellised balconies, through the centre of which the chains passed. This was a saloon or bazaar. The west cottage was the

49 *near right*
Decaying elegance;
kiosk on Brighton
West Pier

50 *far right* With the
pavilion long ago
destroyed by fire and
the shore end
buildings recently
replaced, these small
turrets are the only
original erections
remaining on
Hunstanton Pier

51 Constrasting
ironwork; strong
simplicity below
deck, ornate delicacy
above; Llandudno
Pier

pier-master's house. These cottages survived until 1929. To the west of them a private esplanade originally led to the entrance, where the Palace Pier now touches the land, but when the Aquarium was opened in 1871 the entrance was moved to the shore end of the pier, and two octagonal toll-houses were built just south of the cottages, under the cliff.

During the 73 years' life the pier survived many of the terrible storms that strike the south coast of England from time to time. But on the night of 5 December 1896 the whole structure disappeared from view almost in a single moment. Only the first pair of towers remained visible but not erect the next day and was soon after demolished.

Strangely enough the Chain Pier at Brighton did not inspire the construction of whole series of piers of suspension bridge type. But it did have one echo over fifty years later in the small pier at Seaview near Ryde in the Isle of Wight. This was a miniature version with four uprights and a triangular-shaped platform at the sea end. It was built in 1880 and lasted until after the second world war. It was in fact the first pier to be listed by the then Ministry of Town and Country

52 Folkestone, typical
of the period, 1887

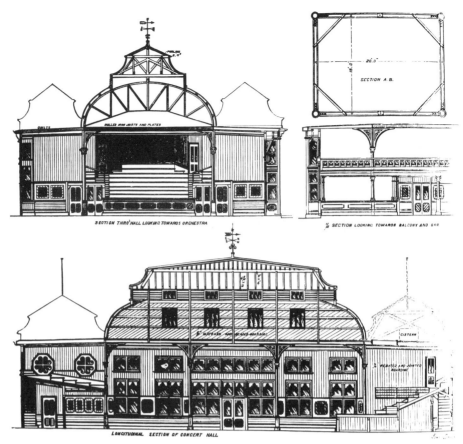

SECTION THRO' HALL LOOKING TOWARDS ORCHESTRA

½ SECTION LOOKING TOWARDS BALCONY AND BAR

LONGITUDINAL SECTION OF CONCERT HALL

Planning (now the Department of the Environment) when the first Ryde list was
made. But it was already in a poor state, had been condemned as unsafe and was
soon afterwards demolished.

The main period of pier building was, as we have seen, in the 1860s. But two
examples appeared in the decade before: Margate (1853–6) and Southport (1859–60).
Neither was an outstanding construction, but both had some importance as trend-
setters. Margate was the earliest pier designed by Eugenius Birch and also had a
technological significance. But it has been so much extended and altered since
then that little of Birch's original work can now be seen. Southport, which was
designed by James Brunlees, is usually called the first true pleasure pier. But it
lacks any romantic feeling because the sea has receded so far at Southport that
most of the pier only passes over land—which is a contradiction in terms of what
a pier is supposed to be.

From 1860 onwards piers came to be built in two forms. The first comprised
those which were only platforms above the water, with no buildings on them.
The second consisted of those which had some kind of superstructure on them. The
first category seemed to concentrate largely on their length, such as Southend
(which, as rebuilt in 1888, is the longest ever constructed), Herne Bay, Saltburn
and Hunstanton. Their qualities lay in what could be seen from them and not in
their own outline. The best surviving pier without buildings on it is Clevedon in
Somerset. This has been listed as Grade II*. It was designed by R. J. Ward &
J. W. Grover and built in 1867–8. It stands a good deal higher above the sea than
most piers and derives its grace from the supports on which it rests and the lacy
ironwork in the spandrels between them which has something of the gossamer
nature of a spider's web. Unfortunately this pier is now breached and condemned

53 The elegant
St Leonards built,
rather unusually, at
the shore end, 1888.
The structure was
demolished in 1951

as unsafe.

It should perhaps be emphasised that, even in the case of piers with some kind of buildings on them, such buildings were in the early years only of the lightest possible nature: kiosks or the like. Theatres and concert halls were almost invariably an addition of a more sophisticated and popular period at the turn of the century. Curiously enough the two piers which today give the best idea of what the early piers of the 1860s looked like originally with only kiosks on them, were actually later buildings. Both are in Wales: Llandudno (1876) and Bangor Garth (1896). They will be described later in their chronological order.

In the realm of piers with superstructures the most important name, as mentioned in Chapter 4, is that of Eugenius Birch (1818–84). His pier connection was massive as he designed no less than fourteen—Margate, Brighton West, Blackpool North, Aberystwyth, Deal, Hornsea, Lytham, Plymouth, New Brighton, Eastbourne, Scarborough, Weston-super-Mare Birnbeck, Hastings and Bournemouth.

Birch's first pier was Margate. But much the most significant of his works and the one which influenced the design of all later piers, both his own and others, was the West Pier at Brighton. This was begun in 1863 and opened on 6 October 1866.

When it came to making a design for the pier Eugenius Birch must have noticed that two buildings in Brighton showed the influence of the Orient. The Chain Pier, which was still unscathed in the sixties, spoke of Egypt, the Royal Pavilion of farther east still. He decided to take inspiration from the latter. This was a bold decision to make at that time for, in the national context, the Pavilion was then very much out of fashion. In Brighton itself the Pavilion has never really gone out of fashion, either then or since, as was shown in 1850 when the inhabitants

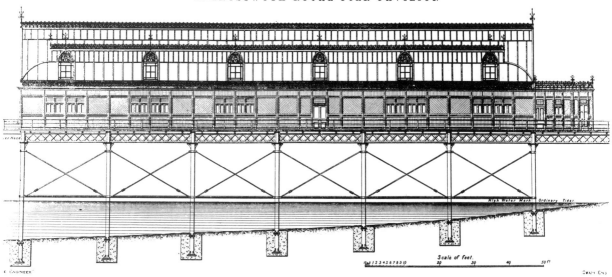

demanded that the building should be purchased for the use of the town, which was upheld by a public poll.

The influence of the Pavilion on the new pier was both general and particular. For instance the seats which run the whole length of the pier and also operate as a handrail or balustrade are divided into sections by higher ribs having much the same form as some of the dragons in the music-room of the Pavilion. On the handrail at intervals are placed standard lamps, originally lit by gas, round which twirl other kinds of music-room dragons. The general influence pervades the superstructures. These compromise two rectangular kiosks at the shore end, two octagonal ones one third of the way down the pier and four more octagonal ones at the angles of the sea platform. All have wide projecting eaves with iron scroll brackets and pendants and either low French pavilion roofs or inset galleries with turrets above. All these pavilions are of vaguely oriental conception. This was to set the style and tone for almost all piers to come. It might almost be called the special seaside-fun oriental style and had an echo as recently as the Festival of Britain, 1951.

How very little sophistication the West Pier and the other early piers possessed can be seen in the original primitive arrangements for bands. The only band-stand, if that is not too grand a word, was a circular platform at the sea end raised about a foot above the ground and large enough to hold approximately ten musicians.

It was similar to the stand provided for rustic musicians in the Skansen Park at Stockholm before 1939 or such as would be suitable for *platzmusik* in one of the smaller villages of the Austrian Tyrol. A little later on, when the centre of the pier was widened, a covered but still tiny band-stand was made there which looked like another kiosk, polygonal in plan and with a pointed roof. This did duty until the first world war.

Since 1866 the West Pier has undergone many alterations and additions, but these have not spoiled or vulgarised its silouette, as has happened in the case of so many piers. The windscreen down the centre was added in 1890 but has an attractive little domed or barrel-shaped roof with a series of pinnacles that continue the oriental feeling of the pier. At the same time the sea end was widened and a large pavilion built, which was converted into a theatre in 1903. This rather

obscures the four original kiosks of the end platform between which is is placed. It has arcaded sides with covered balconies over and a tiered roof. It pays lip-service to the oriental tradition in a rather silly little turret on the sky-line at the north end.

The concert hall halfway down the pier was not built until 1916. It is striking that such a non-essential building could be erected in the middle of the war. This addition, though a fine building in itself, does not follow the oriental theme of the rest of the pier. It is in a style somewhat reminiscent of the Grand Palais in the Champs Elysées at Paris. It is of polygonal plan with segmental headed windows. But the chief feature is the roof line. This forms a sort of elongated elliptical dome divided by sections of balustrading that rise vertically across the dome and by lateral lantern lights, the whole enclosed in a parapet topped by urns and scrolled cartouches.

Even with these additions the pier remains one of the three best in Great Britain and equalled by only one other in England. It is graded by the Department of the Environment II*.

The concert hall of this pier having been built in 1916, it is interesting to note that the last burst of life shown by the pier related also to the first world war. When the film of *Oh! What a lovely war!* was made the theatre of the pier became the headquarters of the film-makers. The pier was specially redecorated for the purpose and many of the shots were taken on it. Soon afterwards the axe fell. The sea end was closed as being unsafe in 1971, the shore end in 1975. Its future remains a great question-mark. Of the several piers now under threat of demolition this and Bangor Garth would be the greatest loss.

Of Eugenius Birch's other piers, Blackpool North is probably the best (Grade II). It has four kiosks not dissimilar to those of the Brighton West Pier and good iron-work. Weston-super-Mare Birnbeck (Grade II) is unusual in that the pier constitutes a bridge to an island with an odd tail-portion at an angle forming a landing stage. It has the same clever combination of balustrade with seats as at Brighton; so does Eastbourne (Grade II). But very little of Birch's work however survives in the latter. Halfway down its length there are two elegant classical pavilions or games-saloons facing each other which have the air of stable blocks

57 Weston-super-
Mare Grand Pier; the
pavilion dates from
the early 1930s, and
now houses an
amusement arcade

58 The new theatre
pavilion on Southsea
Pier, rebuilt in
traditional style
after the 1974 fire

of an eighteenth-century country house. These were added in 1901. Birch's other surviving piers have been much altered, Hastings vulgarised out of recognition and several demolished altogether.

Of the piers not designed by Eugenius Birch, the first of any significance was Llandudno (Grade II*). This was the work of Brunlees and McKerrow and was built in 1876. It has had a pavilion added at its sea end. But, apart from this, it still presents very much the appearance of one of the early piers sustaining nothing but a series of very elegant but fairly large octagonal kiosks with wide overhanging eaves. The pier is unusual in that it does not, as most piers do, follow a single straight line but takes a turn of nearly 45 degrees about a third of the way down its length. It plays a singularly important role in Llandudno as this resort retains more of its social nineteenth-century atmosphere than perhaps any seaside town in Great Britain, with its donkeys on the sands, a concert-party, brass band, steamer trips to the Isle of Man and other homely features of holidays by the sea which more sophisticated places have lost in our time.

The 1880s saw the inauguration of a pier that did not follow in the oriental tradition initiated by Eugenius Birch. This was Skegness, which was designed by Clarke and Pickwell and built in 1880–1. It was approached by an extremely lumpy Gothic archway flanked by smaller kiosks in matching style. These would not have been out of place on the edge of a nineteenth-century cemetery. But their heaviness combined with the association of Gothic with ecclesiastical or at least serious architecture made them singularly unsuitable for a pier. It is not surprising that they were replaced in 1937. But it was unfortunate that the replacement did not even have the merit of being odd. This was an exceedingly commonplace amusement arcade, which in turn vanished in 1970.

St Anne's (1885) was roughly contemporary with Skegness. Its original outline was not of great interest. But in 1904 it followed the example of Skegness in erecting an entrance gateway of non-conforming type. This was of almost residential character and built in red brick and timbering like an unsuccessful imitation of a Norman Shaw house.

At the same time at the other end of the pier, however, the normal pier tradition was followed, and a Moorish pavilion was added, which has since been a casualty in a fire.

The 1890s brought the erection of two remarkably fine piers which still survive intact. The first of these is Bangor Garth, which was designed by J. J. Webster in 1896. As it has never had any pavilions or other buildings added to it, this pier gives the best idea of what most of the early piers looked like when they were purely promenade-places above the sea. It has only a series of very elegant polygonal kiosks with steeply curving roofs with a slight suggestion of Chinese influence about them. This pier has been graded II* and is one of the three best surviving in the country. But unfortunately it is one of those that have been condemned and is closed to the public. Its disappearance would be a major loss in the context not only of Wales but of Great Britain as a whole—in fact little, if any, less than that of Brighton West Pier.

The other major pier of the 1890s is Brighton Palace. The Marine Palace and Pier Company was formed in 1886 in order to build it and to replace the Chain Pier. Work started in 1891 but the company ran into financial difficulties. Work had to be suspended for some years, and a hoarding was put up to screen from view the part of the pier that had already been erected. It was only the intervention of the philanthropist John (later Sir John) Howard that enabled work to be resumed, and the pier was opened on 20 May 1899. The designer was R. St. George Moore. This pier is the most oriental of all in design, which is not surprising in view of the fact that it stands almost opposite to the Royal Pavilion and could probably be seen from it before the trees in Old Steine grew up. The oriental note was struck from the outset as it was entered originally through three lacy segmental arches surmounted by cresting which from the first were lit by electric light. These are repeated all down the length of the pier, but the three at the entrance were replaced in 1930 by a more conventional jubilee-type clock-tower, when the promenade was set back for road widening. Where the pier widens there is a series of kiosks, some octagonal with wide eaves projecting on scroll brackets and other rectangular with the eaves treated like pelmets. The balustrades and all the iron work are of most delicate scroll patterns.

Many additions have been made to the Palace Pier, but nearly all are in matching style. Unlike most piers, it was always intended from the start that the Palace Pier should have a pavilion or concert hall at its sea end. But work on this was not begun until the pier had been opened. It was completed in 1901. This had a single-

storeyed Moorish arcade on all four sides, which is still there. Within the rectangle was a concert hall with a barrel-vaulted glass roof. Viewed from the north and south this had something of the appearance of one of the aisles of the Crystal Palace. At its angles and rising one storey above the arcades were four turrets of oriental design. Each was surmounted by a small dome at each of its own corners and by a larger dome in the centre. These were originally gilded. In 1939 they were blacked out and since then have only been restored to the half mourning of silver grey. In 1911 the concert hall was converted into a theatre and a glass café added on the first floor at the north end. This and other later alterations have badly mutilated the building. For instance the turret at the north-west corner now lacks its domes.

At the same time as the theatre was brought into existence a new concert hall (now the Palace of Fun) was built halfway down the pier. This has no oriental flavour but is a pleasant building with circular windows and a central dome. Its most interesting feature is perhaps the enriched cornice and spandrels of the circular section within the dome. The windscreen down the pier had been added five years previously in 1906. This deliberately sustains the oriental note as it is surmounted at intervals by pairs of half-arches which face away from each other like horns. These and the cresting of the triple arches which are repeated at intervals down the pier are really its most prominent feature.

The Palace Pier (Grade II*) was the grandest pier to be built. It may not have been superior to the West Pier as the latter was first designed, but since the alteration to the latter it certainly is so now.

Somerset Maugham says somewhere that every composition should have a beginning, a middle and an end. Brighton fulfils this recommendation as far as

60 Herne Bay Pier pavilion following fire in 1970

61 Southsea Pier, 1974. Firemen confined the fire to the pavilion, which was later rebuilt

piers are concerned. The Chain Pier was the beginning of the fashion; the West Pier was the middle of it and the Palace Pier virtually the end. Very few new piers were built after 1900. Most of these were in East Anglia and none were of any significance. Cromer (Grade II) was perhaps the best; Fleetwood (1910) was the last of the prewar vintage. Deal alone has a new pier dating from after the war. This was built in 1954–7 of reinforced concrete. It is typical of the mid-twentieth century to prefer durability to grace.

But many piers received alterations or additions after 1890–1900. By the Edwardian period seaside resorts were much more crowded than they had been in the 1860s. As we have seen, the visitors were more sophisticated and not content simply with promenading above the sea. They expected entertainments such as concerts, theatrical performances, and, later, amusement arcades. Pavilions for these had therefore to be built. A few of these which have disappeared are worth mentioning.

The first was St Leonards, where the whole pier has vanished. This was designed by R. St. George Moore, the architect of the Brighton Palace Pier, and built between 1888 and 1892. It was a charming affair with a colonnade and roof-balustrade in the Chinese manner which was reminiscent of the Chinese dairy at Woburn.

The other two examples were added to pre-existing piers of no special distinction. Clacton pier dated originally from 1870–1 and was designed by P. Bruff. It was widened and a pavilion added at the sea end by Kinipple & Jaffrey in 1890–3. The pavilion had a polygonal plan with two curved ends. These ends had wide overhanging eaves and a balcony at first floor level, both of which were edged with pelmet treatment in the classical Regency manner.

Nearly contemporary was the pavilion added by G. Croydon Marks to the pier at Aberystwyth which had been designed by Eugenius Birch in 1864-5. This was opened by Queen Alexandra as Princess of Wales on 26 July 1896. It stood near the shore end and comprised three aisles surmounted by domed glass roofs. The elevations were ribbed and scrolled in what can only be described as a mixture of Gothic, classical and oriental styles. Apart from these three examples and the first pavilion at the Brighton Palace Pier, which was part of its original design, it can hardly be said that any of the new pavilions added anything to the looks of the piers on which they were imposed. At best some did not spoil them. But many amended the silhouette for the worse. In the years following the war amusement arcades often replaced concert halls. Most of these were of a solid and far from lacy construction. Commercialism became the dominant note and some piers were vulgarised to a sad degree.

The second war dealt them a crippling blow: for the most part they were closed; some were breached; none were maintained—with disastrous effect, as is at present being discovered. The older ones have now celebrated their centenary. Several have been condemned and are presenting their communities with the gravest problem to restore them. At least two (Brighton West and Bangor Garth) should undoubtedly be saved. But it may well be that cast iron just will not stand in the sea for longer than a hundred years without being totally renewed. If this is so and these threatened piers disappear, then a moral should be drawn from it. Brighton Palace Pier is thirty years younger than the West Pier. It is probably the best surviving pier anyhow and certainly will be if Brighton West and Bangor Barth disappear. Except for recent damage by a barge to the north-west end it seems in good condition and is in working order. The twenty-three years left before this pier also celebrates its centenary should be used by the management with the financial help of the municipality and the Government, through the Historic Buildings Council, to see that this pier is henceforward so well maintained that it does not become a new prize problem piece as the twenty-first century dawns.

62 Brighton West and Palace Piers. The latter
is a popular and successful structure, the
former the subject of great conservation
controversy

SIX

PIERS TODAY

There are, in 1976, just over fifty seaside pleasure piers in Britain. In some areas they are few and far between. Those which still exist have survived many hazards, ranging from neglect to military action, and from fire to storms, and their current states vary enormously.

At the one extreme, there are several piers which are structurally sound, are commercially successful and socially popular. They have been developed and improved to keep in step with changing public demands and have been designed to cater for a specific market. Often, they are highly sophisticated entertainment complexes with their facilities put to use throughout the year. Such piers constitute an important or, more accurately, an essential part of their resorts. The three piers at Blackpool are obvious examples. Each has a separate role to play in Blackpool's tourist life. The original North Pier caters for the 'traditional holiday-maker', providing deck chairs on a wide sun deck for sedentary pursuits, and such well established facilities as a theatre and cafe. The other two piers—Central and South—cater for the tripper whose day is more active, with amusement machines and the like sharing space with the traditional entertainments. All three piers are owned by a nationally based firm of entertainment and catering specialists and their popularity and commercial success seem to assure their future.

Other piers are also in use throughout the year for promenading purposes (often with refreshment facilities) with the major entertainment facilities being brought into use in the summer months. In the winter, the piers are used mainly by local residents as promenades and rendez-vous, and by out of season visitors. Bournemouth Pier, for example, provides a pleasant attraction through the year.

There are other piers aimed solely at the summer trade, which are closed in the winter months. Not only are running costs cut, but the structure is also available for overhaul and repair. Llandudno and Swanage are two such piers.

At others, only the extreme shoreward end is in use, often with an amusement arcade or similar facility provided. The rest of the pier is in limited use as a promenade (as at Felixstowe and Southwold) or else completely closed (as at Hunstanton).

At the lowest extreme some piers are completely closed to the public and are threatened with demolition, or at best, with accelerated decay resulting from neglect. Often such piers are the subject of major conservation controversy.

The factors responsible for these varying fortunes are manifold. Old age and decay are obvious threats to piers. Maintenance and repairs have sometimes been neglected in the past, and the resultant decay is now making itself known. The result has been closure or, occasionally, collapse. Renewal of pier insurance has sometimes been refused because of structural condition and closure has followed. On the other hand, careful maintenance has resulted in some piers being in excellent condition. It does not necessarily follow that the younger the pier the better its condition.

Ship collision still claims occasional victims. Brighton Palace Pier was one. Damaged estimated at £100,000 was caused in October 1973 when a barge, used in the demolition of the pier-head landing stage, broke loose in a storm. As pier officials and crowds of sightseers watched anxiously from the promenade, the barge was dashed against the west side of the pier damaging or severing twenty-five piles. The helter skelter tower, crazy maze, hoopla stall and telephone kiosk, along with part of the decking, tumbled into the sea, where they were joined by the men's toilet after it had been cut free by demolition gangs. One corner of the pier theatre was left hanging precariously over the water, and it was fortunate that the barge sank before being swept right through the pier. The only bright spots in the whole proceedings were the queues of people who streamed through the turnstiles at 6p. a head to view the damage at first hand.

Storms had more direct consequences. Saltburn Pier had suffered repeated storm damage from the day it was built, and late in 1974 the pier head and some ten sections of neck were lost in a winter gale. The pier, originally 1,400ft long and later reduced to 1,280ft was thus cut to 1,100ft.

At Teignmouth, heavy seas exposed some of the pile bases on the pier, and truck loads of sand and cement had to be rushed to the resort to ensure the safety of the structure.

Fire continues to claim an alarming number of pier victims. In January 1970, only excellent work by Eastbourne and East Sussex firemen prevented serious damage to Eastbourne Pier after an arsonist fired the theatre at the seaward end. Five months later the pavilion at the shoreward end of Herne Bay Pier was burned down. Southsea South Parade Pier was the scene of a spectacular blaze on 11 June 1974. Filming of the rock opera *Tommy* was in progress when fire broke out in the draperies of the pier theatre, apparently started by a spot light. An orderly evacuation of several hundred actors, actresses, extras and film staff took place, although two or three people, quite unnecessarily, leaped into the sea from the pier head. Firemen arriving at the incident were faced with the perennial problems of pier fires. The building's age and its mode of construction, assisted by sea winds, allowed rapid fire spread, whilst poor access and, ironically, shortage of water, hindered fire fighting operations. However, determined efforts by over 100 firemen prevented damage to the rest of the structure. By the following year, the building was rebuilt and again ready to play its full part in the holiday life of Southsea.

Social and financial factors

The fortunes of piers lie ultimately in the hands of the public. Piers are provided for the enjoyment and entertainment of the the holidaying Briton (and also that of the foreign visitors) and to be successful they must attract the maximum number of people. However, they face strong competition. The trend for overseas

63 *above* Bournemouth Pier in heavy use. Owned by the local authority, the pier is both socially and financially successful

64 *right* Threats to demolish a number of piers have prompted a vigorous response. This has been particularly marked in Brighton and Clevedon

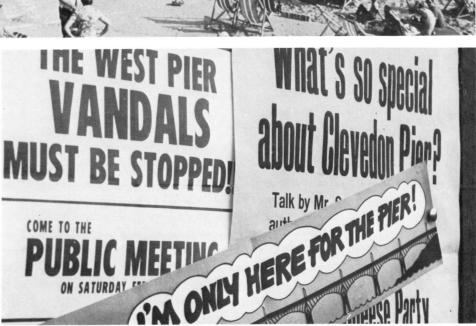

REPAIR AND RECONSTRUCTION

65 *right* Repairs in hand at Brighton Palace Pier following damage in 1973 when a drifting barge sliced through the structure

holidays, so popular in the late sixties, continued into the seventies. A fortnight in Majorca could be as cheap as two weeks in Margate. Benidorm was as accessible as Bognor. The vast increase in car ownership made holidaymakers in this country more mobile than they had ever been. There was some relief for the domestic resorts (and hence their piers) in the mid-seventies. Unsettled political situations in holiday areas and, more important, increasing holiday costs coupled with a higher cost of living forced many people to re-discover holiday areas nearer home. A number of piers were improved and modernised to provide the facilities and the allure expected by tourists of the seventies. The old railway-type refreshment rooms and drab paintwork and bare floors gave way to well-equipped restaurants carpeted throughout; night clubs and lounge bars with cabaret replaced the old theatres and music halls and well-equipped shops superseded the kiosks. In the few cases where this has happened, the decisions to change have been amply justified by the popularity of the structures.

Many piers, however, continued much as before, either because pier operators thought the traditional approach still suitable for clients (as indeed it was in some places) or because the large capital expenditure required for such drastic changes was not available. Some piers compromised between the two extremes.

A number of piers typify those in use in the 1970s, and their recent histories demonstrate the trials and tribulations of current pier ownership. Firstly, there are those with an assured future, and a brief selection taken from the gazetteer will serve to illustrate their different problems and the means taken to solve them.

Clacton Pier, Essex

In contrast to many piers, Clacton must represent pier utilisation and exploitation at its utmost. A vast collection of amusement and entertainment hardware covers much of the deck of this structure. Facilities provided include a big dipper, helter-skelter, cable car, merrygoround and aquarium (with dolphins and sea lions) in addition to the traditional pier theatre. Clacton's near neighbour, Walton Pier, duplicates many of these attractions, and also has a pier railway.

These piers are undoubtedly brash and commercialised, bearing little resemblance to the Victorian structures from which they grew. They are, however, both successful and popular and thus fulfilling the hopes of their builders.

Southend Pier, Essex

The decreasing receipts and increasing expenditure associated with this pier resulted in some urgent financial study by its owners, the local authority, around 1970. Large capital expenditure was required, but the council felt unable to provide this from the rates, and an alternative solution was found. Basically, the pier remained the property of the local authority, but the running of the entertainment facilities and amenities on the pier was entrusted to a catering and leisure firm for an annual rental of £25,000. The council would receive a percentage of all profits above a set figure.

The lessees spent some £250,000 on the pier to provide a new restaurant and amusement arcade, pub, cocktail bar, and night club. There were advantages to both pier and town from those facilities. The traditional draw of the pier itself brought a large number of potential clients to the new facilities, and, with no rival establishments nearby, they enjoyed good business. The novelty of the situation must also have been an enticement to patrons. Furthermore, the noise and other nuisances associated with late night entertainment were safely away from the town.

These arrangements did not totally solve the problem, however. Within a few years there were moves to demolish the pier, but the local authority decided in the mid-1970s to spend some £3m on restoration over a fifteen-year period.

Herne Bay Pier, Kent

Across the mouth of the Thames, Herne Bay Pier had been closed in 1968 when an insurance survey revealed some of the supports to be in a critical condition. The pier was threatened and the local authority, as the pier owners, debated its possible future. Three basic schemes—the restoration of the whole pier, the repair and re-opening of a 'medium length' pier or the retention of only the landward end with pavilion—were examined and the last named plan decided upon, with possible extension of the work later. The destruction of the pavilion in 1970 further complicated the matter.

Proposals and counter-proposals, including the demolition of the structure, resulted in a plan to erect a large sports pavilion on the landward end of the pier. A variety of sports facilities were included in the design, including a sauna bath, two squash courts, a bar, committee rooms and a main hall available for roller skating, badminton or hockey. The building is currently under construction, but controversy still rages over the fate of the remainder of the pier. Local anglers have campaigned for its restoration in order to establish Herne Bay as a major sea angling centre but no firm decision has yet been taken on the future of the long pier neck.

Bournemouth Pier, Dorset

One particularly successful and popular pier is that at Bournemouth. It is available throughout the year for strollers and anglers and a pier-head café is open. The pier theatre really comes into its own in the summer when it is used for variety shows which maintain the high standard of pier entertainment over the years. The pier is owned by the local authority and regularly makes a profit for the rates, as well as acting as a major attraction in this resort.

Paignton Pier, Devon

Success is not only confined to large commercialised piers. Paignton, to the west of

Bournemouth and one of the smaller structures, is the base for cruises along the coast and up the River Dart. A motor cruiser now plies the route where, before the war, a paddle steamer would have been used. There are small scale amusements such as go karts, amusement machines, and a children's playground, as well as refreshment facilities.

Grand Pier, Weston, Avon

Some forty years younger than the Birnbeck Pier at Weston-super-Mare, the Grand Pier has a large pavilion at the pier head housing dozens of amusement machines, dodgem cars, bingo and other such diversions. There are a number of shops at the pier entrance selling refreshments and souvenirs, and the rest of the extensive deck area is given over to promenading areas and seating.

Birnbeck Pier, Avon

This pier is undergoing a period of transition. The islet forming the centre section of the pier supports a number of buildings which have been put to a variety of uses. A carriage museum and banqueting hall, complemented by souvenir shops and a pier railway (a fleet of carriages towed by ex-London taxi cabs) have given way to a 'cash and carry' shop. Present plans are for the erection of an hotel on the islet to provide a useful amenity in an unusual setting. The main part of the pier, forming as it does a vehicular and pedestrian bridge linking the islet to the shore, would form an essential part of such a project.

Mumbles Pier, South Glamorgan

Like several others, this pier has its financial backing concentrated at the shore-ward end, where an amusement arcade, helter skelter, ballroom and hotel are located. The rest of the pier is open and maintained, but it has little money-making potential, and the future of this section may not be assured. Nonetheless, the pier constitutes a useful and popular attraction in this South Wales resort.

Threatened piers

These are some of the successful piers, and those for which, by one means or another, a future is assured. There are a number of piers, however, which are in danger of demolition or partial closure. Generally speaking, privately owned piers should make a profit or else they are altered to a suitable standard or even closed and demolished. Where the owner is a local authority, however, the matter is not so simple. Piers may be regarded as amenities, in the same way as parks, and there is a case for involving public money in their upkeep. Bournemouth Pier, as stated, runs at a profit but a publicly owned pier running at a loss would be acceptable in some cases. In other instances, piers have been offered to local authorities only after the private owner has failed to make a viable enterprise of them.

It is ultimately finance which determines the fate of threatened piers, and unfortunately the 1970s have seen many examples of piers in danger.

Saltburn Pier, Cleveland

A publicly owned structure with a black future, Saltburn Pier aroused national interest when the 1974 storm damage made demolition seem likely. The local authority applied for listed building consent to demolish the whole structure as, subsequent to the storm, it was suggested that repairs to the entire pier would cost some £350,000. An alternative scheme to retain only the landward section was costed at £60,000. A temporary reprieve has been granted.

Bangor Pier, Gwynedd

Bangor's publicly owned pier was closed in 1971 because the structure was unsafe. Local conservationists rallied to its support, and they had an ally in the necessity to obtain an Act of Parliament to demolish the structure. Repair estimates at the time of closure were in the region of £500,000, but in June 1975, the local authority decided the leave the pier exactly as it was for a three-year period. The structure has been temporarily reprieved but decay continues unabated.

St Annes Pier, Lancashire

This pier, privately owned, seemed to have an assured future until 1974, when an early morning blaze destroyed the Floral Hall on the pier head. Estimates for replacing the building were in the region of £1m and further repairs were also considered necessary to the pier. Consequently, demolition was decided on. Like other piers, the structure was listed by the Department of the Environment as being of architectural or historic significance and alterations could only be effected with government consent. As soon as the demolition plans become known, local conservationists campaigned for the pier's reprieve and petitions with over 6,000 signatures, together with individual letters of protest, were sent by residents to the local authority. The council, after much discussion, refused consent to demolish the last 150ft of the 1,200ft pier on the grounds that: (a) loss of the buildings, jetty and pier head would constitute a serious loss of amenity for both residents and visitors; and (b) the appearance and character of the pier would be seriously affected.

Revised schemes for the end of the pier are currently being prepared.

Other piers are currently closed to the public and may be in danger. New Brighton and Shanklin are two examples. In two places, however, the threats to piers have been the source for major controversy.

Clevedon Pier, Avon

On 17 October 1970, part of Clevedon pier collapsed under routine load testing. Clevedon and its pier had not acquired the trappings of a brash twentieth-century resort, but remained a quiet middle class seaside town. The pier's popularity had gradually declined. In the 1950s, the consulting engineers to the local authority (the pier's owners) had instituted a programme of testing the pier every other year to issue a certificate of structural soundness to obtain insurance cover. Temporary water tanks were erected on the pier to simulate the loading agreed with the then Ministry of Transport. If the pier withstood these loads, it passed the test. In 1970 it did not.

The result was to unite all sections of Clevedon society in a desire to restore the pier. A preservation society was formed to raise money for the purpose and plans were drawn up to effect repairs which did not detract from the structure's appearance. Unfortunately, fund raising continually lagged behind repair estimates, whilst further deterioration took place on the derelict structure. Following local government reorganisation in 1974, a thorough inspection of the pier was carried out by the local authority engineer. This survey revealed that in addition to the main restoration work, further repairs and maintenance of a general nature were required. In fact, the £75,000 quoted for reinstatement shortly after the collapse was totally inadequate four years later, some £380,000 then being considered a reasonable estimate for total repair.

Certainly the pier's plight attracted national as well as local interest. Clevedon's role as part of a proposed conservation area, and the pier's status as a listed building, raised hopes of a state grant to help meet restoration costs. The local authority, as the pier's owners, would have to meet maintenance costs. Unfortunately, the prospects of recouping some of the cost through money-making facilities on the pier seem limited, and it would serve primarily as an amenity. At the present time, the pier is still closed and restoration has yet to take place.

Brighton West Pier, East Sussex
Even more controversy resulted from the proposal to demolish Brighton West Pier, and a different approach to conservation was adopted. Totally different circumstances permit the possibilities of running this pier as a commercial venture.

Brighton West is owned by the West Pier Company, 97% of its share capital having been acquired by AVP Industries in 1965. At that time plans were announced for developing the pier as a comprehensive holiday and conference centre, open throughout the year and in all weathers. Nothing, however, was done until 1969 when an application was made for consent to demolish the main part of the pier. This was agreed for the southern (seaward) part of the pier only.

Under the West Pier Act, the owners were obliged to effect repairs to the pier. Various attempts to give the threatened section of the pier away failed since with the structure went the awesome responsibility of repairing and maintaining it. Also, under the terms of the Act, the local authority could serve a repairs notice on the owners; if they failed to act, the council could carry out repairs and recover the cost from the pier owners. There was, however, the chance that the company would then go into liquidation (its only asset being the pier) leaving the local ratepayers to foot the bill.

Various attempts were made to resolve the situation and save the pier by means of financial agreements between the owners and the local authority. All such attempts failed and the repair notice was rescinded because of the expense and difficulty of preparing it. The pier now stands in a sea of uncertainty, but it has some powerful allies.

The 'We Want the West Pier' campaign was formed to campaign for the pier's purchase by the local authority and its revival as a major attraction in the resort. The campaigners' careful and detailed study revealed two important facts. Firstly, the pier could be repaired and restored for a much more reasonable sum than had been supposed. Secondly, with the provision of varied and imaginative attractions on the pier, it could be a profitable enterprise, eventually contributing to the income of the town.

Various engineering surveys in the past had quoted widely varying sums for the repair work. The campaign reports gave detailed consideration to two alternative repair schemes costing, respectively, £1,325,000 and £600,000, the annual maintenance costs being higher for the second scheme. To pay for these repairs, the campaign suggested various attractions and facilities on the pier which would cater for three main sources of income: delegates and their guests attending conferences in the town; foreign visitors; and domestic tourists and local residents. Again, alternative schemes were suggested but both consisted basically of providing a children's entertainment area at the pier entrance, with traditional pier facilities in kiosks. A concert hall and continental café at the midpoint of the pier and a pavilion with night club and restaurant at the pier head completed the scheme.

Investigations carried out by the preservationists exploded the myth that piers are not profitable. Clacton, Weston Grand, Teignmouth and Hastings Piers (all privately owned) made handsome profits in the early 1970s, in addition to the obvious commercial successes of Blackpool. Detailed study of probable income from a reopened West Pier indicated all loans being repaid within ten years.

The pier campaign was successful in preventing the planned demolition. The local council are currently evaluating the studies carried out on the pier's future and their decision is awaited by the pier's supporters throughout the country.

Engineering and architecture

If the West Pier campaign focussed attention on the financial complexities of maintaining piers, it also demonstrated the engineering difficulties encountered in pier repair. These problems arise not so much in actually carrying out the work but more in estimating what needs to be done.

In the Brighton case, a 1969 report on the whole pier quoted £203,000 for repairs, whilst a later estimate (January 1971) for repairs to the southern end only was £650,000; repairs to the northern end were considered to cost another £100,000. A December 1973 estimate for southern end repairs was £906,000. Assessments made by the pier campaign engineers in December 1974 were £842,000 (southern end) or £941,000 plus maintenance for minimum repairs to the entire structure (including new buildings). Maintenance costs were estimated at £42,700 p.a. for 20 years, and thereafter at £30,000 p.a.

Figures are quoted to give some idea of the magnitude of costs and are swiftly outdated. Much of the discrepancy between estimates was due to inflation. Nonetheless the differences are partly due to different approaches to the problem.

Estimates are necessarily based on an inspection of the structure. In practice, a sample of members (say 20 per cent) may be studied and the results extrapolated to cover the entire pier. However, decay and damage does not, by any means, follow a consistent pattern. Substructural members may be of varying dimensions, different ages and various materials (often as a result of extensions or previous repair work) with consequently differing degrees of decay. Recent research has also shown that, given a virgin structure, corrosion will vary from one part to another. In fact, components sheltering from direct exposure may corrode more severely because of the combination of still air, condensation and build up of salt deposits. Consequently, proposed repairs, and their cost, may not all be necessary.

The projected life of repairs will also affect the cost. It may be decided to replace wholesale all suspect components to guarantee a life of a certain number of years, or it may be decided only to carry out 'emergency' repairs as and when the occasion arises. Coupled with this is the question of maintenance. A comprehensive programme would obviate the need for frequent replacement of components, but little or no maintenance resulting in emergency repairs periodically might be more attractive in the short term.

The means of repair and materials utilised also affect repair estimates. Traditional repairs to girderwork generally consist of cutting out corroded steelwork and welding matching sections in its place. Such work is usually carried out by erecting scaffolding beneath the pier and lowering materials from above. This is satisfactory for occasional repairs but time consuming and expensive for large scale works. Repairs to columns generally consist of encasing damaged bases in concrete, or else placing cylindrical 'splints' around columns further up.

These methods have the advantage of retaining much of the appearance of a pier and are suitable for general maintenance work. At Brighton West, however, the large scale works required resulted in proposals to depart from traditional methods. Repair schemes envisage the use of hollow box girders as trusses and beams, with concreted repairs to existing columns, or their replacement by square steel stanchions. Braces and ties would be replaced with hollow steel sections and solid round bars respectively. A carefully planned repair programme working from deck level through gaps left by removing deck timbers would result in speed of working and less expense.

One problem, however, which has occurred in other cases is the reluctance of contractors to carry out repair work for a fixed price. A satisfactory solution has normally been the reimbursement of a contractor's costs plus a percentage of overheads and profit. Publicly owned piers have often benefited from the services of direct labour operating under the direction of local authority engineers.

Examples of pier repair work in the 1970s are many, varying from minor work to large scale reconstruction. In the former category girder replacement at such piers as Eastbourne and Southsea are two examples. The winter of 1975–6 saw routine maintenance at Bournemouth including blasting and repainting of sub-structural components and repairs to decking. At Birnbeck Pier, in the same period, repairs were undertaken to main girders and decking. Among the timber piers, Great Yarmouth Wellington Pier was partially reconstructed in 1972–3. Timber piles, steelwork and decking on the end section were replaced.

Herne Bay Pier falls in to the latter category. The destruction of the pavilion in 1970 eventually led to plans for its replacement. Examination of the existing columns indicated that the concrete pads on which they were mounted would require enlarging; this was carried out at successive tides by concreting round the bases. The structure supported by these members was of steel columns atop the cast iron columns, carrying steel roof trusses.

Civil engineering has taken great steps forward in the last hundred years. Engineers and contractors have the use of sound theory, computing techniques, and plant and machinery unheard of when Birch and his contemporaries built their piers. Nonetheless, pier work in the 1970s still needs something of the skill and ingenuity, not to mention tenacity, displayed by the Victorian pier builders. The validity of the idea of civil engineering being the attempts to combat or harness the forces of nature for the use of man is amply borne out by the generations of piers which have been built out into Britain's seas in the last 170 years.

The future

The long-term future of piers is uncertain. It is unlikely that any new pleasure piers will be built—sheer cost would probably rule this out. It is the existing piers which will carry the pier tradition into the future. Some will undoubtedly survive because of their commercial success, but the fate of the less successful structures, and those currently requiring large scale repair, is uncertain.

Many regard piers as anachronisms, examples of Victorian frivolity which have long outlived their usefulness. There are others, however, who see piers as essential and irreplaceable facets of the British seaside. The anglers who fish from the pier, the trippers who use their amusement and refreshment facilities, the strollers who patrol their decks, and the observers who merely enjoy the visual spectacle of piers, should all be concerned with the future of piers.

Piers are examples of engineering excellence, of architectural frivolity and of novel responses to social needs. Even today, a walk on the pier has great attractions, and new generations of holidaymakers and seaside residents continue to derive great pleasure from these structures. As tastes have changed so piers have, generally, changed with them and in modernised and altered forms continue to fulfil the role of pleasure and amusement emporia. They have suffered from many hazards and their continuing existence is indicative of their undeniable place in the holiday tradition. It would indeed be a pity if indifference and callousness were to hasten their demise. But not until the last one has finally succumbed to the onslaught of wind and sea will the pier story come to an end.

68 Bangor Pier 1975, closed whilst the owners and local residents debate its future

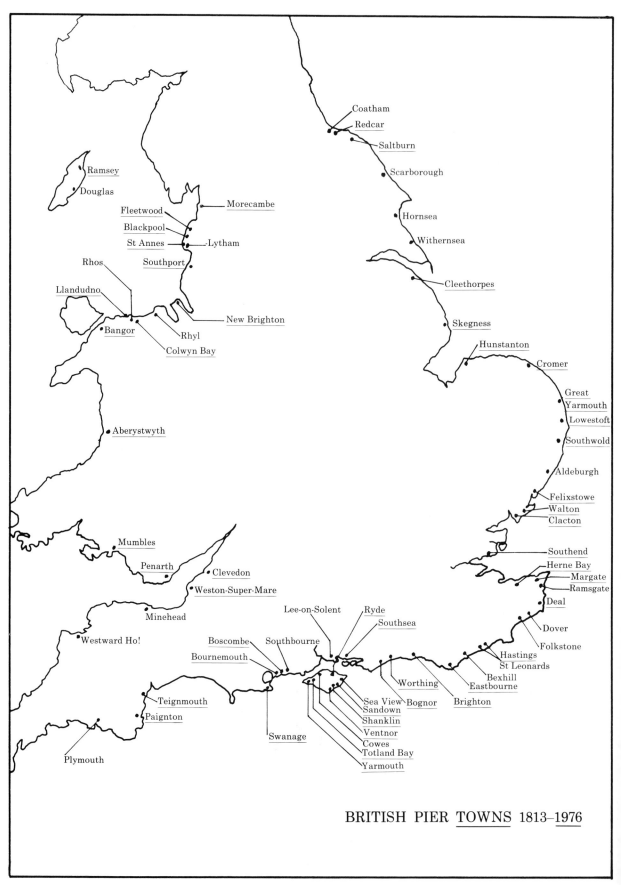

Ramsey
Douglas

Coatham
Redcar
Saltburn
Scarborough
Hornsea
Withernsea
Cleethorpes
Skegness

Morecambe
Fleetwood
Blackpool
St Annes — Lytham
Southport

Rhos
Llandudno
Bangor
Rhyl
Colwyn Bay
New Brighton

Hunstanton
Cromer
Great Yarmouth
Lowestoft
Southwold
Aldeburgh
Felixstowe
Walton
Clacton
Southend
Herne Bay
Margate
Ramsgate
Deal
Dover
Folkstone

Aberystwyth

Mumbles
Penarth
Clevedon
Weston-Super-Mare
Minehead
Westward Ho!

Lee-on-Solent   Ryde
Southsea
Boscombe   Southbourne
Bournemouth
Hastings
St Leonards
Bexhill
Eastbourne
Brighton
Worthing   Bognor
Sea View
Sandown
Shanklin
Ventnor
Cowes
Totland Bay
Yarmouth

Teignmouth
Paignton
Swanage

Plymouth

BRITISH PIER TOWNS 1813–1976

69 West Pier, Brighton, 1965. The future of this pier is still uncertain.

APPENDIX 1 Chronological list of piers

Pier	Built	Engineer	Contractor	Demolished	Length	Remarks
RYDE	1813–14				1,250	Extended in 1815, 1824 and 1833; head built 1842 and extended 1859. Separate tramway and railway piers later added.
BRIGHTON CHAIN	1822–3	S. Brown	MacIntosh /Brown	1896	1,134	Damaged by storm in 1833 and 1836, but subsequently repaired. Destroyed by storm in 1896 whilst awaiting demolition.
SOUTHEND (1)	1829–30			1887	1,800	1835–46: extended to 1¼ miles and new head to designs of J. Simpson. Replaced by new pier in 1887.
WALTON-ON-NAZE (1)	1830				330	
HERNE BAY (1)	1831–2	T. Telford		1870	3,613	Extensive repairs and replacement of wooden piles by cast iron after 1839.
DEAL (1)	1838	J. Rennie		1857	250	Proposed 445ft length never completed.
GREAT YARMOUTH WELLINGTON	1853–4	P. Ashcroft			700	Extensively reconstructed in 1900–3 following purchase by local authority, including provision of theatre. Large scale renewal of components in 1971.
MARGATE	1853–6	E. Birch	Bastow/direct			First pier to incorporate screw piles. 1875–8: pier head extended (with pavilion) by G. G. Page. 1893: further extensions.
GREAT YARMOUTH BRITANNIA (1)	1857–8	A. W. Morant	G. Allen	1899	700	Breached by ship collision and shortened 1859 and 1868. Demolished and replaced by present pier.
SOUTHPORT	1859–60	J. Brunlees	W. & J. Galloway		3,600	Extended and improved in 1864 and 1868. Damaged by storm 1889, and successive pavilions burned down in 1897, 1933 and 1959.
BOURNE-MOUTH (1)	1859–61	G. Rennie	D. Thornbury	1876	1,000	Wooden structure damaged by decay and storm. Replaced by temporary jetty in 1877 and new pier in 1878–80.
WORTHING	1861–2	R. Rawlinson	J. Cliff		960	Enlarged, and pavilion added, in 1887–9 to designs of J. Mansergh. Widened 1913–14 with new buildings in 1926, 1935 and 1959.
BLACKPOOL NORTH	1862–3	E. Birch	R. Laidlaw		1,410	Extended and improved in 1874 and 1903 with shops and pavilion, with further improvements in 1932 and 1937. Fire damaged in 1921 and 1938, the pier was last rebuilt in 1966.
BRIGHTON WEST	1863–6	E. Birch	R. Laidlaw		1,115	1893: extended, and pavilion added. 1916: concert hall provided. 1932: a new top deck entrance built.
DEAL (2)	1863–4	E. Birch	R. Laidlaw	1954	1,100	Badly damaged during last war, and demolished shortly afterwards.
RYDE VICTORIA	c 1864			1922	900	Never fully completed.

Pier	Built	Engineer	Contractor	Demolished	Length	Remarks
ABERYSTWYTH	1864–5	E. Birch	J. E. Dowson		800	1866: 100 ft of pier destroyed; reopened 1872. Pavilion and substructure at shore end added by Marks in 1895–6. Badly damaged by storm 1938.
BOGNOR	1864–5	J. W. Wilson	J. E. Dowson		1,000	Greatly enlarged with addition of buildings in 1910–11. Pavilion at seaward end collapsed into sea March 1965.
LYTHAM	1864–5	E. Birch	R. Laidlaw	1960	900	Floral hall erected c1895. Pier breached by ship 1903, and pavilion destroyed by fire 1927. Closed 1938.
TEIGNMOUTH	1865–7	J. W. Wilson			600	Enlarged c. 1887 and buildings at seaward end completed 1890.
EASTBOURNE	1866–72	E. Birch	J. E. Dowson Head Wrightson		1,000	1877: shoreward end destroyed by storm and rebuilt. 1888: pierhead theatre built, with second theatre in 1899–1901. Further buildings erected in 1901, 1912, 1925 and 1951.
SCARBOROUGH	1866–9	E. Birch	J. E. Dowson Head Wrightson	1905	1,000	1883: damaged by ship collision. 1889: pavilion built at shore end, but pier destroyed by storm January 1905.
NEW BRIGHTON	1866–7	E. Birch	J. E. Dowson			Extensively rebuilt 1928–30 by Wallasey Borough Engineer, and new pavilion erected.
BLACKPOOL CENTRAL	1867–8	J. I. Mawson	R. Laidlaw		1,518	Entrance altered and improved in 1877 and 1903. New theatre and entertainments complex opened 1968, and destroyed by fire 1973.
CLEVEDON	1867–8	R. J. Ward J. W. Grover	Hamilton's Windsor Ironworks		842	1891: cast iron building erected on enlarged and realigned head. 1970: two spans of neck collapsed under test loading.
RHYL	1867	J. Brunlees	R. Laidlaw	1972	2,355	Ship collision in 1883 and burning of pavilion in 1901 culminated in closure in 1913. Repaired and reopened in 1930, but closed in 1966 with final length of only 330 ft.
WESTON-SUPER-MARE BIRNBECK	1867	E. Birch	Isca Iron Co.		1,350 plus island	Two stage pier incorporating an islet in its length.
SALTBURN	1868–70				1,400	1900: storm damage reduced length to 1,280 ft, and further damage occurred in 1924 and 1930. Theatre built 1925. Severe damage 1974.
DOUGLAS	1869	J. Dixon	J. Dixon	1896	1,000	Pier dismantled, removed and rebuilt at Rhos-on-Sea 1896.
HASTINGS	1869–72	E. Birch	R. Laidlaw		910	Pavilion burned down 1917, and subsequently replaced.
WESTWARD HO!	1870–3	J. W. Wilson	Gooch, J. Abbott	1875	493	The projected six month construction took three years, and instability of the structure led to its demolition two years after completion.
TOTLAND BAY	c1870	S. H. & S. W. Yockney				
HUNSTANTON	1870–1	J. W. Wilson			830	Pavilion destroyed by fire 1939.
CLACTON	1871	P. S. Bruff				1890–3: length increased to 1,180 ft, and structure considerably widened and improved. New head (by Hocking to designs of Kinipple & Jaffrey) and pavilion (Murdoch & Cameron) provided.
REDCAR	1871–3	J. E. & A. Dowson	Head Wrightson		1,300	1880/85/97: damaged by ship collision. 1898: head burned down. 1909: pavilion erected on widened shoreward end, and extended 1928. War and storm damage resulted in present truncated pier.
CLEETHORPES	1872–5		Head Wrightson		1,200	1903: pierhead pavilion destroyed by fire. Replaced by shops and midpier pavilion erected. 1938: new entrance provided. Following breaching in 1940, seaward end of pier demolished.
VENTNOR	1872					1885: rebuilt and lengthened following damage, and repaired and enlarged in period 1889–1913. 1951–5; rebuilt to designs of B. Phelps by Wall Bros.
MORECAMBE CENTRAL	c1872				912	1896: pier strengthened. 1897: two pavilions built, both destroyed by fire 1933. New pavilion and ballroom built 1935–6.

Pier	Built	Engineer	Contractor	Demolished	Length	Remarks
COATHAM	1873–5			1899	1,800	1874: breached by ship whilst still under construction. Further breach in 1898, and outermost section demolished.
HERNE BAY (2)	1873	Wilkinson & Smith	Direct labour		320	1884: pavilion and theatre erected at shoreward end. Pier incorporated in third pier of 1896–9.
WITHERNSEA	1875–8	T. Cargill	J. O. Gardiner	c1900	1,196	1880: 200 ft removed by ship collision and later replaced. Further damage resulted in only 300 ft being usable in 1890 and 50 ft by 1900, despite extensive reconstruction.
ALDEBURGH	1876–8	T. Cargill	G. W. Hutchinson	NK	561	
LLANDUDNO	1876–8	Brunlees & McKerrow	J. Dixon		2,295	
YARMOUTH	1876		J. Denham		665	
BOURNEMOUTH	1878–80	E. Birch	Bergheim		838	Building erected on pier in 1885, followed by extensions in 1894 and 1909. Breached 1940, repaired and reopened 1946. Head reconstructed 1950, with new theatre and substructure in 1960.
PAIGNTON	1878–9	G. S. Bridgeman	J. Harris			1919: pavilion destroyed by fire. Breached in 1940 and later repaired.
SANDOWN	1878				360	1895: extended to 875ft. 1934: new pavilion built. 1971–3: new entertainments centre erected on rebuilt shoreward end.
SOUTHSEA SOUTH PARADE	1878	G. Rake	Head Wrightson		600	1904: badly damaged by fire. Reopened in 1908 following rebuilding by Bevis, Yelf and Thorne to designs of G. E. Smith. 1967: theatre removed. 1974: pavilion damaged by fire and subsequently rebuilt.
HORNSEA	1879	E. Birch	Bergheim	1910		1880: ship collision destroyed pavilion and part of neck. Pier completely derelict by 1897.
RAMSGATE	1879–81	H. Robinson	Head Wrightson	1930	550	Pier badly damaged by fire and ship collision c.1917, and by mine in 1918.
SEA VIEW	1880	F. Caws	Various	1951	1,000	
SKEGNESS	1880–1	Clarke & Pickwell	Head Wrightson		1,802	1919; pier breached by ship collision and subsequently repaired.
RAMSEY	1882		Head Wrightson		2,150	
PLYMOUTH	1884	E. Birch		1953	465	1890–1: pavilion erected. c.1940: damaged by enemy action.
LEE-ON-SOLENT	1885–8	Galbraith & Church	J. Robinson	1958	750	
ST. ANNES-ON-SEA	1885					1974: pavilion destroyed by fire.
FOLKESTONE	1887–8	M. N. Ridley G. Chatterton	Heenan & Froude	1954	683	1890: floating landing stage erected at head. 1940: breached. 1945: pavilion destroyed by fire, pier badly damaged.
SOUTHEND (2)	1887–8	J. Brunlees	Arrol Bros		6,600	1898; pier extended and new head built; designers J. Brunlees and J. W. Barry, contractors Murdoch and Cameron. 1908: upper deck built by Borough Engineer/C. Wall. 1929: Prince George extension by Borough Engineer/Peter Lind.
BOSCOMBE	1888–9	A. Smith	E. Howell		600	1927–8: pier head rebuilt. 1940: breached. 1958–60: neck rebuilt and roller skating rink opened 1962.
ST LEONARDS	1888–91	R. St. G. Moore	Head Wrightson	1951	950	Pavilion designed by F. H. Humphreys. 1909: extensively altered. 1940–51: breached, damaged by military action, storm and fire.
SOUTHBOURNE	1888	A. Smith	E. Howell	1907	300	1900–1: almost entirely destroyed by storms.
BRIGHTON PALACE	1891–1901	R. St. G. Moore	A. Mayoh		1,760	1901: theatre opened. 1910: shore end pavilion built. 1938: pier extended. 1973: head badly damaged by drifting vessel, and subsequently repaired.
SHANKLIN	1891				1,200	1927: pavilion destroyed by fire.

PIER CASUALTIES

70 *above* Dover Pier,
1910, demolished 1927
after a life of less than
forty years

71 *right* Ramsgate
Pier in 1915,
demolished 1930
following damage in
the 1914–18 War

72 *below* Lytham Pier,
1918, demolished 1960.
The pier was one of
Birch's early designs

Pier	Built	Engineer	Contractor	Demolished	Length	Remarks
BLACKPOOL SOUTH	1892–3	T. P. Worthington	J. Butler		429	Pavilion by J. D. Harker. Major fires in 1958 and 1964 destroyed amusement arcade, shops, cafe and 'Rainbow Theatre' (later rebuilt).
DOVER	1893	J. J. Webster	A. Thorne	1927	900	Ship collision 1893 and storm 1894 damaged pier and led to closure. Reopened 1895, pavilion (by W. J. Adcock) opened 1899. Closed 1926.
MORECAMBE WEST END	1893–6	J. Harker	Mayoh & Haley Widnes Foundry Co		1,800	Extensive gale damage in 1903 and 1927 reduced pier to present 900ft. Pavilion destroyed by fire 1917.
SWANAGE	1893				642	Extensively repaired 1928.
PENARTH	1894	H. F. Edwards	J. & A. Mayoh		650	1927–8: landing stage and pavilion erected. 1931: pier badly damaged by fire, and by vessel in 1947. Subsequently repaired.
WALTON-ON-NAZE (2)	1895		J. Cochrane		800	Length increased in stages to 2,600ft.
BANGOR	1896	J. J. Webster	A. Thorne		1,500	1914: badly damaged by drifting vessel, subsequently repaired. 1971: closed.
HERNE BAY (3)	1896–9	E. Matheson	Head Wrightson		3,787	1910: enlarged by Widnes Foundry Co. to designs of Matheson. Pavilion by Sands, Pattinson to designs of Waldram, Moscrop, Young, Glanfield. 1928: pavilion of 1884 and 1970: pavilion of 1910 burned.
RHOS-ON-SEA	1896	J. Dixon		1954		Douglas Pier rebuilt.
MUMBLES	1897–8	W. S. Marsh	Mayoh & Haley Widnes Foundry Co.		835	1940: breached. 1956: new landing stage built. 1966: amusement arcade built.
BEXHILL	1898	Mayoh & Haley				Abandoned before completion.
COLWYN BAY	1899	Maynall & Littlewood	Widnes Foundry Co.		220	1922: pavilion burned down. 1923: pier extended and new pavilion built. 1933: pavilion burned down and replaced 1934. Repair work to pier in 1954 and 1964 onwards.
CROMER	1900–1	Douglass & Arnott	A. Thorne		500	Damaged by storm in 1949 and 1953. Subsequently repaired.
GREAT YARMOUTH BRITANNIA (2)	1900–2	Mayoh & Haley	Widnes Foundry Co.			1909: pavilion destroyed by fire. Rebuilt but again burned down 1914. 1928: ballroom erected, destroyed by fire in 1932. Pier further damaged by fire in 1954.
SOUTHWOLD	1900	W. Jeffery	A. Fasey		800	1936: pavilion erected at shore end. 1955: outer section of pier washed away and length reduced to 370ft.
COWES	1901–2	R. E. Cooper	A. Thorne	1961	170	1951: pavilion demolished.
MINEHEAD	1901	J. J. Webster		1940	700	
LOWESTOFT	1902–3	D. Fox	S. J. Trueman		760	1912: T-head extended; washed away with part of pier in 1962.
WESTON-SUPER-MARE GRAND	1903–4	P. Munroe	Mayoh & Haley		1,000	
FELIXSTOWE	1905		Rogers Bros.		2,640	Substantially rebuilt and shortened.
FLEETWOOD	1910	G. T. Lumb			492	
DEAL (3)	1954–7	Sir W. Halcrow & Partners	Concrete Piling		1,026	Only completely new post-war pier.

APPENDIX 2 Gazetteer of surviving piers

CLEVELAND

1 Redcar 1871–3/J. E. & A. Dowson/
Head Wrightson   SE 608 252
Only the shore end pavilion (1909) and
some 70ft (21.4m) remain of this pier,
which originally had a length of
1,300ft (396.5m). Cantilevered girders
springing from cast iron columns
support the wooden decking. The
shoreward end was widened by the
use of new columns to support the
pavilion portal trusses. Breaching
during the war and subsequent
damage resulted in the present
structure, which is publicly owned.

2 Saltburn   1868–70/NK/NK
NZ 666 217
This is the last survivor of a range of
piers which once graced the Yorkshire
coast. Trestles of braced cast iron
columns carrying wooden deck beams
carried the pier 1,400ft (426.7m) out
into the sea, and the pier terminated
in a landing stage for steamers and
pleasure craft. Storm damage had
reduced the pier to 1,280ft (390.2m) by
late 1900, and following damage by
ship collision in 1924 part of the neck
was rebuilt in steel. The pier was also
breached by the military during the
last war and subsequently repaired.
Unfortunately, a storm in late 1974
destroyed the head and some ten
sections of neck. Thus, the pier now
consists of two sections, i.e. the
damaged seaward section of 1930, and
the inner 650ft (196m) which is mainly
original, although the wooden deck
beams have been replaced by mild
steel components.
The pier is also publicly owned.

HUMBERSIDE

3 Cleethorpes   1875/NK/Head
Wrightson   TA 318 090
The original length of this iron piled
pier was 1,200ft (365.8m), its width 20ft
(6.1m), and the pier had a concert hall
at the seaward end. This concert hall
was destroyed by a fire in 1903, and a
new building was erected halfway
along the pier. A cafe and shops were
built on the site of the older building.
The elevated link to the adjoining Pier
Gardens was removed in the late 1930s,
and the pier breached in the early
part of the 1939–45 war. The isolated
seaward section was subsequently
demolished, and the pier thus reduced
in length to its present 335ft (102.1m).
The pavilion has been extended and
modernised and comprises a 600 seat
concert hall, cafe and bar. The local
authority are the owners.

LINCOLNSHIRE

4 Skegness   1880–1/Clarke &
Pickwell/Head Wrightson   TF 572 634
Forty-four entries were submitted in
response to the Skegness Pier
Company's offer of a £50 premium for
designs for their proposed pier. The
winning design featured a large pier
head pavilion and pairs of shelters
on widenings of the neck. The
substructural work was typical of the
period, with screw piles and cast iron
columns supporting lattice girders. In
addition, comprehensive bracing
secured the columns to a line of centre
piles. Since the original construction,
the pier has been modernised to
include a recent leisure centre and bar
at the shoreward end. The pier is still
owned by the original promoters.

73 Saltburn Pier, before the destruction of the pavilion

NORFOLK

5 Hunstanton 1870–1/J. W. Wilson/ NK TF 671 409
This tall, graceful pier, is of a similar design to Wilson's ill-fated Westward Ho! Pier, and incorporates a large amount of timber on top of iron piled columns. Its original length of 830ft (253m) has been shortened to 675ft (205.7m), whilst the pier-head pavilion was destroyed by fire in 1939. An amusement arcade was erected at the shoreward end in 1962–3, and although the pier is currently closed to the public, it is the outstanding East Anglian pier. It greatly enhances the appeal of this small resort. The Hunstanton Pier Co. are the owners.

6 Cromer 1900–1/Douglass & Arnott/A. Thorne TG 219 424
Cromer Pier was built for the Cromer Protection Commissioners at the end of the pier building era. It was short and wide, with a consequently high earning potential and low maintenance costs. Within its 500ft (152.4m) length are a pavilion and various amusement facilities. The relative youth of the pier is indicated by the extensive use of steel girders atop the usual cast iron columns and wrought iron piles. Despite repeated storm damage, the pier has been regularly maintained by the local authority and is currently in good condition.

7 Gt. Yarmouth (*Britannia*) 1900–2/ Mayoh & Haley/Widnes Foundry Company TG 531 077
This unprepossessing steel and wood structure replaced a wooden pier of 1857–8. Although originally provided with an imposing pavilion by Boulton and Paul of Norwich, the pier is now dominated by later amusement buildings, and the 810ft (246.9m) structure may simply be regarded as a platform to support these facilities.

8 Great Yarmouth (*Wellington*) 1853–4/P. Ashcroft/NK TG 531 067
Ashcroft's original wooden pier was rebuilt in 1900–3 following its purchase by the local authority. The pier was widened at the shoreward end by means of steel girders atop wooden piles, and the present pier theatre erected. There is a short promenade deck seaward of the theatre. The local authority are the owners.

SUFFOLK

9 Lowestoft (*Claremont*) 1902–3/ D. Fox/S. J. Trueman TM 545 919
Originally of timber, some 760ft (231.7m) long, and with a T-head, the pier is now 40ft (12.2m) shorter, and reinforced concrete has replaced some of the timber. The shoreward end is enclosed to form a sun-lounge with the usual facilities, whilst the remainder of the pier is given over to fishermen and the occasional stroller. Like the piers at Great Yarmouth, Claremont Pier could not really be said to be an attractive structure.

10 Southwold 1900/W. Jeffrey/ A. Fasey TM 512 767
Another rather unprepossessing East Anglian pier, Southwold was originally built entirely of wood and was some 800ft (244m) long. However, war and storm damage has resulted in the truncation of the pier to its present length of 370ft (112.8m). The short promenade deck is almost bare of accessories and the structure's main raison d'être appears to be the two storey amusement arcade (of interwar construction) at the shoreward end.

11 Felixstowe 1905/NK/Rogers Bros TM 300 341
Many of the remarks on Southwold Pier could well be applied to this one. The large amusement arcade dominates the pier, whilst what remains of the neck is uncluttered, almost devoid of decoration, and mainly of reinforced concrete. Pier Amusements (Felixstowe) currently own the structure.

ESSEX

12 Walton-on-the-Naze 1895/NK/ J. Cochrane TM 254 215

Erected to replace a jetty of 1830 by the Walton-on-Naze Pier & Hotel Co., the pier was originally 800ft (244m) in length. Successive lengthenings and widenings resulted in the present length of 2,600ft (792.5m), and its extensive collection of bowling alley, amusement arcades, fairground equipment, etc., all housed in a large, hangar-like building at the shoreward end. About the only thing the pier lacks is visual appeal. The New Walton Pier Co. owns the pier.

13 Clacton   1870–1/P. Bruff/NK
TM 177 145
Bruff's memel pier was considerably extended in 1890-3 to the designs of Kinipple and Jaffrey. The length was increased to 1,180ft (360m) by contractor William Hocking, a pine polygonal pier head was provided and Murdoch and Cameron erected a most impressive iron pavilion. The pier has been enlarged and extended since then. It must now be the most extensively utilised pier in the country. The shoreward half of the pier is covered with every conceivable entertainment facility and resembles a fairground. The pier head is also provided with various attractions. This is the concept of a pleasure pier taken to the ultimate degree.

14 Southend   1829–30/NK/NK
TQ 884 851
This, as the pier head proudly proclaims, is the longest pier in the world. Originally 1,800ft (548.6m) long, it was extended to nearly 7,000ft (2,133.6m) by James Simpson in 1835–46. The wooden pier was extensively rebuilt in 1888 by Arrol Bros to the designs of James Brunlees, and in 1898 the pier head and extension were completed. Further extensions were undertaken in 1927. The pier is served by its own electric railway which is carried alongside the promenade deck. The pier itself is very complex and encloses a considerable area of water within its various arms. Part of the pier head itself is three tier with the various levels accessible according to the state of the tide. Besides the various amusement facilities there are coastguard and lifeboat stations.

KENT

15 Herne Bay   1896–9/E. Matheson/
Head Wrightson   TR 173 683
At 3,787ft (1,154.3m) this is one of the longest in the country, and

incorporates a much shorter pier by Wilkinson & Smith of 1873. The pier consists of iron piles carrying steel girders and wooden decking with a pitch pine landing stage. The Grand Pier Pavilion of 1910 built 300ft (91.5m) offshore was replaced in 1974 by a massive rectangular box-like structure, and the pier beyond closed to the public. Its future is at present uncertain; the local authority are the owners.

16 Margate   1853–6/E. Birch/
S. Bastow   TF 353 712
This pier is of interest for two reasons. It was the first pier designed by the doyen of pier engineers, Eugenius Birch, and it was the first application in England of Mitchell's screw piles, which became an essential feature of most subsequent piers. It was extended in 1875–8 by George Gordon Page at a cost of £34,457, and featured an octagonal head. Further additions were made in 1893 and 1900. There is an amusement arcade at the shore end, a boat station midway along the pier neck, and a small pavilion at the seaward end. It is owned by the Margate Pier and Harbour Co. 1812.

17 Deal   1954–7/Sir W. Halcrow & Partners/Concrete Piling Ltd.
TR 378 527
This is the only completely new post-war pier in the country, and is built entirely of reinforced concrete—this should ensure its survival for many years to come. The pier is 1,026ft (312.7m) long and replaced Birch's pier of 1863–4 which was damaged beyond repair during the war. This is a local authority pier.

EAST SUSSEX

18 Hastings   1869–72/E. Birch/
R. Laidlaw   TQ 811 091
A typical Birch pier, with a length of 910ft (277.4m), this structure varied in width from 45ft to 190ft (13.7m to 57m). Cast iron columns on screw piles support a lattice girder framework and wooden decking. There are covered walkways, two theatres, and numerous amusement arcades including a zoo. The Hastings Pier Co. owns the structure.

19 Eastbourne   1866–72/E. Birch/
J. E. Dowson and Head Wrightson
TV 618 989
After Brighton West, this is probably the finest of Eugenius Birch's piers. Cast iron screw piles supporting an

iron and wooden frame 1,000ft (304.8m) long now carry a variety of buildings. The shoreward end was destroyed by a storm in 1877, and was rebuilt somewhat higher than the remainder of the pier, to which it is connected by a ramp. Theatres were built in 1888 and 1899, and two games saloons added midway along the pier in 1901. The high standard of finish of these buildings and the ironwork of the appointments combine to make this pier most attractive. Eastbourne is one of several piers owned by Trust Houses Forte.

20 Brighton (*Palace*) 1891–1901/ R. St. G. Moore/ A. Mayoh TQ 314 038
This is certainly one of the most famous of all piers. When construction started in 1891, it was the intention that the 'Brighton Marine Palace and Pier' should replace Samuel Brown's illustrious Chain Pier, which, having been built in 1822–3, was reaching the end of its life. In the event, the Chain Pier was destroyed by a storm in 1896—three years before the Palace Pier was opened. The pier is 1,760ft (536.5m) with a minimum neck width of 45ft (13·7m) and a head width of 189ft (57.6m). Cast iron screw piles grouped in clusters of six along the neck supported a framework of lattice girders on which are mounted rolled steel joists. The famous theatre was opened in 1901, but is closed at present following collision damage incurred in 1973. The pavilion at the shore end with its impressive metal structure was erected in 1910. Permission is now being sought to alter the pier head. It would indeed be a pity if this prime example of the 'architecture of amusement' were to lose any of its great visual charm. The owners are the Palace Pier Co.

21 Brighton (*West*) 1863–6/E. Birch/ R. Laidlaw TQ 303 041
This is Eugenius Birch's finest pier and, perhaps, considering its impact, the most important pier ever built. It was in the van of iron pier development, and set the standard for a generation of piers to follow. Built with an overall length of 1,115ft (309m) and terminating in a head 310ft × 140ft (94.5m × 42.7m), the pier originally had only a few shelters atop it. However, the pier head was extended and a pavilion opened in 1893, a concert hall was opened in 1916, and a new top deck entrance provided in 1932. The pier's future is currently at

the centre of a major conservation controversy in Brighton. It is owned by A.V.P. Industries.

WEST SUSSEX

22 Worthing 1860–2/R. Rawlinson/ J. Cliff TQ 150 023
Originally rather a small, bare structure, the pier has a length of 960ft (292.6m). In 1887–9, the width of the first 750ft (229m) was increased from 15ft to 30ft (4.6m to 9.2m), and a pavilion was erected on the widened head (105ft—32m) in place of earlier buildings, all to the designs of James Mansergh. In 1913–14, the pier was further widened at the shore end and at the mid-point of the neck, and a pavilion was erected in 1935 after the previous building was destroyed by fire, and the amusement arcade midway along the pier was opened in 1959. The pier is owned by the local authority.

23 Bognor 1864–5/J. W. Wilson/ J. E. Dowson SZ 934 988
This was Wilson's first pier (others followed at Hunstanton, Teignmouth and Westward Ho!) and was originally a rather graceful 1,000ft (304.8m) structure, almost devoid of buildings. In 1910–11, the width of the pier was increased to 80ft (24.4m) for almost a third of its length and a large pavilion erected at the shoreward end. In March 1965, the pavilion at the sea-ward end collapsed into the sea, and since then more of the neck has gone. Some of the decking has been removed, and now only the shoreward end with its rather unattractive amusement building is open to the public. This pier is privately owned.

HAMPSHIRE

24 Southsea (*South Parade*) 1878–9/ G. Rake/Head Wrightson SZ 652 981
This pier was damaged by fire in 1904 and much of the current pier dates from the rebuilding completed in 1908 by Thorne, Bevis, and Yelf to the designs of G. E. Smith. Another serious fire in 1974 destroyed the pier theatre, although the pier itself was largely undamaged. The pier is owned by the local authority and has a length of 600ft (182.9m).

ISLE OF WIGHT

25 Ryde Various
SZ 594 929
The original pier on this site was built in timber with a length of 1,250ft

74 Cleethorpes Pier in the 1920s, at the height of its popularity

75 Cleethorpes Pier in the 1970s, now much truncated

76 *below* Clacton Pier; fun fair in the sea

(381m) in 1813–14. It was the first major piled passenger pier and the ancestor of all the piers considered in this survey. The alignment and part of this early pier are incorporated in the promenade section of the existing multiple pier. The pier was extended to 2,250ft (685.8m) by extensions in 1824 and 1833. The present pier, which is owned by British Rail, is a composite of three quite separate trackways: the pedestrian and vehicle deck perpetuating the original pier, the railway track itself, and between them the disused tramway. Each occupies an independent structure. Whilst the pier and pier station have no claim to architectural distinction, the unique transport history gives the pier a distinction which should be recognised.

26 Sandown   1878/NK/NK
SZ 598 840
Built originally to a length of 360ft (109.7m), the pier was extended to 875ft (266.7m) in 1895. The original iron piles have been extensively replaced by reinforced concrete, and much of the girderwork and decking has been renewed. A pavilion was built in 1934, and a new entertainments complex with theatres, bars and shops was opened in 1973. A new concrete landing stage protects the head. All this recent work carried out by the owners—the local authority—has given the pier a new lease of life.

27 Shanklin   1891/NK/NK
SZ 586 814
This is the most attractive of the six surviving Isle of Wight piers, although it is unfortunately closed at the present time. Built for the Shanklin Pier Company, it consists of a girder framework on cast iron columns to a length of 1,200ft (365.8m). The concert pavilion was rebuilt after a fire in 1927, but much of the structure appears to be original.

28 Ventnor   1872/NK/NK
SZ 563 773
The history of this pier has been one of damage, repairs, and extensions. Following war damage, the pier was substantially rebuilt in 1951–5 by Wall Bros to the designs of Basil Phelps. The head is now of reinforced concrete, whilst the neck consists of green heart decking and welded steel beams on cast iron piles. The pier is owned by the local authority.

29 Totland Bay   c. 1870/S. H. & S. W. Yockney/NK   SZ 323 872
This small landing pier consists of a light girder framework on cast iron columns, and carries a minimum of appointments. There is only a small shelter on the head, and a tiny amusement pavilion at the shore end.

30 Yarmouth   1876/NK/J. Denham
SZ 354 989
Built in timber to a length of 700ft (213.4m), this is also merely a landing stage. It has a small shelter on the pierhead, and a brick built pavilion on the shore end.

DORSET

31 Boscombe   1888–9/A. Smith/ E. Howell   SZ 112 911
The original pier was a wood and iron structure 600ft (182.9m) long. It has been entirely rebuilt in reinforced concrete—the head in 1926–7, and the neck in 1958–60. A roller skating rink was opened in 1962. The pier now bears little resemblance to Smith's original design, and is reminiscent of a concrete road bridge in appearance.

32 Bournemouth   1878–80/E. Birch/ Bergheim   SZ 089 907
This pier, one of Birch's last designs, was built for the Bournemouth Improvement Commissioners to replace Rennie's wooden jetty of 1859–61. The original length of 838ft (255.4m) was extended in 1894 and 1909 to give a length of over 1,000ft (304.8m). The lattice girder neck is supported by seven groups of iron screw piles and a landing stage was constructed at a lower level round the exterior of the head and part of the neck. The pier head was reconstructed in 1950, and again in 1960, with the insertion of a new concrete substructure to carry the new pier theatre. Like Boscombe Pier, this structure is owned by the local authority.

33 Swanage   1893/NK/NK
SZ 038 787
This pier is of some interest in that it is built of imported timber in an attempt to overcome the problems which bedevilled the early pier builders. Green heart was used because of its high resistance to decay in a marine environment. However, even this wood suffered some damage and many piles were partially encased in concrete muffs in the 1920s. Now owned by the Swanage Pier Co. the

77 Worthing Pier, a popular and attractive asset to the resort

78 Paignton Pier; small but successful, photographed in 1975

79 *below* Aberystwyth Pier c. 1920. Nearly all of the neck has now disappeared

pier is crescent shaped with a length along the centre line of 642ft (195.7m). It features a two tier head with no amusement facilities and only elementary facilities at the shore end.

DEVON

34 Teignmouth 1865–7/J. W. Wilson/ NK SX 942 727
Only two piers survive in the south-west of England, but neither is very significant. Teignmouth, the better of the two, was a conventional wood and iron structure 600ft (182.9m) long. It is now somewhat shortened and has lost all but its extreme shoreward facilities. It is owned by the Teignmouth Pier Company.

35 Paignton 1878–9/G. S. Bridgeman/J. Harris SX 895 609
This, the second pier in the south-west, is of typical cast iron column/girder construction, and carries a series of long, single storey, unprepossessing amusement arcades. It is owned by the Paignton Pier Company.

AVON

36 Weston-super-Mare (*Grand*) 1903–4/P. Munroe/Mayoh & Haley ST 317 614
This was one of the last of the generation of traditional iron piled pleasure piers, and the brash confidence of its siting in the centre of the seafront provides a fitting close to the era. Its long uncluttered neck consists of a lattice girder framework carried on groups of 10 cast iron columns. The pier head supports a huge rectangular pavilion adorned by turrets at its corners while there are numerous kiosks at the shore end. Owned by the Grand Pier Co. Ltd., this pier is a popular attraction at this busy resort.

37 Weston-super-Mare (*Birnbeck*) 1867/E. Birch/Isca Iron Co. ST 308 624
This pier is most unusual in that it incorporates an island in its design. It consists of a section 1,100ft (335.3m) long from the mainland to Birnbeck Island, and a 200ft (61m) landing section beyond the island.
  The former section consists of an iron framework carried on 15 clusters of cast iron columns supporting a timber deck with small projecting bays, ornamental lamps, and decorative seating along its length. The outer section was originally of timber construction on iron piles but is now also metal girder. The island itself, which carries amusement facilities, bars, a restaurant, museum, and a lifeboat station, had to be levelled, stepped and bricked to form a suitable base for these facilities and for promenading. This pier has an unrivalled Victorian charm possibly due to the fine ironwork; it is most attractive. It is owned by the Weston-super-Mare Pier Co., and there are plans to erect an hotel on the island.

38 Clevedon 1867–8/R. J. Ward, J. W. Grover/Hamilton Windsor Ironworks ST 402 719
This is undoubtedly the most graceful pier in the country. It comprises a masonry approach 180ft (54.9m) long and 20ft (6m) wide and eight 100ft (30.5m) spans of wrought iron girders carried on wrought iron columns and screw piles. The upper parts of these columns consist of pairs of Barlow rails riveted back to back and branching out to form arches beneath the girders. The original pier head consisted of 18 double Barlow rail piles braced diagonally and horizontally with wrought iron rods. The pier head was enlarged and realigned in 1891, when a cast iron building was erected. The two outermost spans of the neck collapsed under test loading in 1970, and despite plans prepared by a preservation trust to restore the pier, no repair work on this listed structure has yet been undertaken. The pier now belongs to Woodspring District Council and their Director of Technical Services has prepared a report summarising the current position and the problems of restoration. Although much of the remaining ironwork and decking needs attention, the comparative shortness of the pier—only 842ft (256.6m)— could act in its favour.

SOUTH GLAMORGAN

39 Penarth 1894/H. F. Edwards/ J. & A. Mayoh ST 190 713
Originally, cast iron piles supported wooden decking, but both ends of this pier have been substantially rebuilt in reinforced concrete. A reinforced concrete landing stage (Mouchel & Partners) and a pavilion (Somerset) were erected in 1927–8, but a serious fire in 1931 caused much damage to the decking, buildings and substructure. Ship collision in 1947 resulted in underpinning of cast iron columns, and the placing of new reinforced concrete columns to the design of Wallace Evans and Partners.

The present pavilion is a pleasant structure echoing traditional pier architecture. The pier is owned by the local authority.

40 Mumbles 1897–8/W. S. Marsh/ Mayoh & Haley, Widnes Foundry SS 630 874

The original structure comprising cast iron piles braced with steel rails, steel lattice girders, and pitch pine decking, has been joined by a three tier reinforced concrete landing stage on steel piles at the head. There is also a lifeboat station arm at right angles to the main pier. The amusement facilities are clustered at the shoreward end; these include an amusement arcade, helter-skelter, ballroom and hotel. However, it would appear that whilst these facilities are secure, the remaining portion of the pier has outlived its usefulness.

DYFED

41 Aberystwyth 1864–5/E. Birch/ J. E. Dowson SN 581 818

The main structure of the pier consisted of braced cast iron piles and columns let into well hole beds in solid rock outcrop and concreted in position. The pier was originally 800ft (243.8m) long, but the outer 100ft (30.5m) were washed away in 1866 and reinstated and reopened in 1872. The pavilion at the landward end was designed by G. Croydon Marks and erected by the Bourne Engineering and Electrical Co. in 1895–6 at a cost of £8,000. The pier was badly damaged by a storm in 1938, and today the pier neck is closed to the public. The pavilion appears from the plans to have been originally quite attractive, but it is now clad in clapboard, and is something of an eyesore. However the facilities provided therein include a ballroom, banqueting hall, restaurant and (in the old cinema) bingo and amusement arcades. These may well survive even if the rest of the pier does not.

GWYNEDD

42 Bangor (*Garth*) 1896/ J. J. Webster/A. Thorne SH 584 732

This pier is essentially for landing and promenading; there are no amusement facilities. The length is approximately 1,500ft (457.2m). The width is generally 24ft (7.3m) with increases at braced towers at 250ft (76.3m) intervals in which kiosks and shelters are situated. The pier head widens out to 59ft and then to 99ft (18m and 31m). There are stairways bracketed off columns and girders giving access to an old jetty, the foreshore, and a new landing stage. The construction is of timber planked decking on timber joists carried on steel cross beams. These span on to steel lattice girders which are supported by cast iron columns terminating in screw piles. There is an ornate handrail to the perimeter of the pier, and ornamental lamps. Periodic maintenance and repair work throughout the last forty years failed to arrest decay and in 1971 the pier was closed to the public as engineers considered it unsafe. No firm decision as to its future has yet been taken.

43 Llandudno 1876/Brunlees & McKerrow/J. Dixon SH 784 830

This pier consists of two sections. The main pier is carried on a wrought iron lattice girder framework on cast iron columns and extends 1,234ft (376.2m) to a T-shaped pier head 60ft (18.3m) wide. It is lined with four pairs of kiosks and has three larger octagonal kiosks on the head. At the shore end, an arm of the platform connects the pier with its pavilion some distance along the shore. The pavilion consists of a projecting gable centre portion and recessed wings with apsidal ends and fronted by a verandah. The pier and pavilion are owned by the Llandudno Pier Co.

CLWYD

44 Colwyn Bay 1899–1900/Maynall & Littlewood/Widnes Foundry Co. SH 852 791

Originally 220ft (67m) long with a width from 22ft to 75ft (6.7m to 20.8m) the pier consisted of braced cast iron columns carrying steel girders and thence timber deck members. In 1922 the pavilion was burned down and in the following year the new owners— the local authority—extended the pier to its present length of 475ft (145m) and erected a new pavilion. This was destroyed some 10 years later and a replacement pavilion was opened in 1934. Various repairs have been effected since the war, and the pier is now in use as a pleasure and amusement pier. It is owned by Trust Houses Forte.

MERSEYSIDE

45 New Brighton 1866–7/E. Birch/ J. E. Dowson SJ 312 941

The original structure, built for the

New Brighton Pier Co., consisted of iron piles supporting four lines of longitudinal main girders on which were mounted transverse wrought iron joists and thence the decking. The pier was repaired and rebuilt by the Wallasey Borough Engineer and Surveyor in 1928–30, and a pavilion erected. The pier, still publicly owned, has recently been closed to the public, and should thus be considered to be in some danger.

46 Southport 1859–60/J. Brunlees/ W. & J. Galloway SE 335 176
This was built primarily for promenading, and therefore may be regarded as the first true pleasure pier. For a time it was the longest in the country, extending to 3,600ft (1,098m). It was extended and widened in 1864 and further extended in 1868 to 4,380ft (1,335m). Major fires in 1897, 1933 and 1959, coupled with storm damage and various alterations, resulted in the present pier of 3,650ft (1,112.5m) with modern amusement buildings, an elevated section over the land, and a pier railway. In the pier's construction, the contractors used the jetting principle to sink the piles in the sand, where piles sank under their own weight into sand agitated by jets of water. The three original rows of piles were increased to four during the widening. The pier is currently owned by the local authority.

LANCASHIRE

47 St Anne's 1885/NK/NK
SD 319 286
Built some twenty years after the now demolished Birch pier at neighbouring Lytham, this is a conventional iron column and lattice girder pier. The landing stage at the end of the pier was reconstructed in 1891 and the Moorish Pavilion and bank kiosk added in 1904. The Floral Hall was erected in 1910 and the gable entrance lodge with its imitation timber framing is also twentieth century. The pier is a listed structure, and a recent application 'to demolish the burnt-out Pavilion, damaged Floral Hall, pier substructure and jetty and construction of a new end feature on the seaward side' was refused consent by Fylde District Council in 1975.

48 Blackpool (*South*) 1892–3/T. P. Worthington/J. Butler SD 305 337
This is the newest of the three Blackpool piers. Originally short (429ft—130.8m) and wide, it carried 36 shops, shelters, and a bandstand. The pier pavilion was designed by J. D. Harker. The pier was lengthened considerably, and more entertainment facilities built. Serious fires in 1958 and 1964 destroyed several buildings, but these have been replaced and the pier is now an essential part of this seaside resort.

49 Blackpool (*Central*) 1867–8/ J. I. Mawson/R. Laidlaw SD 305 355
This pier has an overall length of 1,518ft (462.7m), of which 400ft (122m) is jetty. The pier section is carried on iron columns spaced at 60ft (18.3m) centres with wrought iron girders between. The buildings have suffered periodic fire damage and are all relatively modern. The latest victim was the Dixieland Palace built in 1968 and destroyed in 1973. Like the other two piers in the town, this is in daily use and is owned by Trust Houses Forte.

80 New Brighton Pier; currently closed

50 Blackpool (*North*)   1862–3/
E. Birch/R. Laidlaw   SD 305 365
This pier, the earliest and most
attractive of the three at Blackpool,
was one of Birch's first designs.
1,410ft (429.7m) long, the pier consists
of plate girders spanning 60ft between
clusters of cast and wrought iron
piles. The head was enlarged in 1874
to carry a pavilion, restaurant and
shops, and in 1903 a theatre was built.
Despite the usual alterations following
fire damage and occasioned by
changes in taste, the pier retains
much of its original flavour. The neck
is relatively uncluttered and has two
pairs of early kiosks.

51 Fleetwood   1910/G. T. Lumb/NK
SD 336 486
Built for the Fleetwood Victoria Pier
Co., the pier is relatively short at
492ft (150m). It is carried on iron
columns and is used as a landing jetty.

Amusement facilities are concentrated
at the shoreward end.

52 Morecambe (*West End*)   1893–6/
J. Harker/NK   SD 424 638
This pier was originally 1,800ft
(548.6m) long and of conventional
iron construction. After severe storm
damage it was reduced in length to its
present 900ft (274.3m) in 1927. The
buildings are all relatively modern,
the early pavilion having been
destroyed by fire in 1917. The structure
is owned by its promoters, the West
End Pier Co.

53 Morecambe (*Central*)   c. 1870/
NK/NK   SD 434 646
This 912ft (278m) long pier is notable
for its massive pierhead. The early
pavilions were destroyed by fire and
replaced in 1935–6 by a large ballroom
and pier theatre. The pier is owned by
the Central Pier Company.

81 Bognor Pier, 1976

INDEX